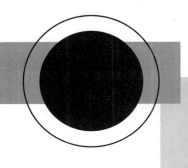

Chip Design for Submicron VLSI:
CMOS Layout and Simulation

John P. Uyemura

Georgia Institute of Technology

THOMSON

Australia   Canada   Mexico   Singapore   Spain   United Kingdom   United States

Chip Design for Submicron VLSI: CMOS Layout and Simulation

by John P. Uyemura

Associate Vice-President and Editorial Director:
Evelyn Veitch

Publisher:
Bill Stenquist

Sales and Marketing Manager:
John More

Developmental Editor:
Kamilah Reid Burrell

Production Services:
RPK Editorial Services

Copy Editor:
Shelly Gerger-Knechtl

Proofreader:
Pat Daly

Indexer:
Melba Uyemura/RPK Editorial Services

Production Manager:
Renate McCloy

Creative Director:
Angela Cluer

Interior Design:
Carmela Pereira

Cover Design:
Katherine Strain

Compositor:
PreTEX, Inc.

Printer:
Transcontinental Printing

Cover Photo:
Yalcin Alper Eken,
Atlanta,Georgia

North America
Nelson
1120 Birchmount Road
Toronto, Ontario M1K 5G4
Canada

Asia
Thomson Learning
5 Shenton Way #01-01
UIC Building
Singapore 068808

Australia/New Zealand
Thomson Learning
102 Dodds Street
Southbank, Victoria
Australia 3006

Europe/Middle East/Africa
Thomson Learning
High Holborn House
50/51 Bedford Row
London WC1R 4LR
United Kingdom

Latin America
Thomson Learning
Seneca, 53
Colonia Polanco
11560 Mexico D.F.
Mexico

Spain
Paraninfo
Calle/Magallanes, 25
28015 Madrid, Spain

Dedication

This book is dedicated to the author, my husband, father of our children,

John Paul Uyemura

for his never-ending and unconditional love and support during our many years together.
I love you John. We love you Daddy.

植　村

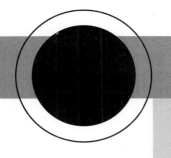

Contents

Chapter 7   *MOSFET Modeling with SPICE 134*

Chapter 8   *CMOS Logic Gates—Design and Layout 157*

Chapter 9   *Standard Cell Design—Layouts and Wiring 198*

Chapter 10   *Storage Elements—Design and Layout 221*

Chapter 15   *Digital System Design 1—The Dsch Program 307*

Chapter 16   *Digital System Design 2—Design Flow Examples 335*

Chapter 17   *Capacitors and Inductors—On-Chip Passive Elements 357*

Chapter 18   *Analog CMOS Circuits—Layout Basics 370*

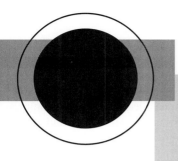

Preface

This book is about the physical design, layout, and simulation of CMOS integrated circuits. It is written around a very powerful CAD program called **Microwind** that runs on PCs that have the Windows operating system. The interface is very friendly, and the program is both educational and useful for designing CMOS chips. When coupled with its companion program **Dsch** (also on the CD), it provides an automated design environment where one can design a logic schematic and translate it in to a CMOS circuit with a few easy steps.

The book centers on physical design as a central theme and is not a comprehensive treatment of VLSI or CMOS circuit design. While it is almost entirely self-contained, most readers will find it useful to consult more general books on chip design. References are provided at the end of most chapters.

● ● ● ● ● ● ● ● ● ● ● ● ● ● ●

Audience

No prior knowledge of CMOS, physical design, or integrated circuit layout is assumed. The book has been written to make it accessible to a wide range of readers. All of the fundamental concepts needed to learn physical design are discussed when they are introduced. The presentation does use basic concepts from electric circuits and digital logic, and it also assumes some basic skills in operating a PC.

In a college level curriculum, the book can be used with a junior or senior level course in VLSI or CMOS integrated circuits. Since the discussion employs many examples, most students in electrical or computer engineering will find it good for self-study.

● ● ● ● ● ● ● ● ● ● ● ● ● ● ● ●● ●●●

Motivation

In June, 2002, I received an email from Professor Etienne Sicard of INSA, Toulouse, France. He told me about a CMOS layout program that he had written for PCs and asked if I was interested in seeing it. I responded positively and received a CD and an instruction manual a few weeks later.

When I first launched the program called Microwind, I found myself in a standard-looking layout editor environment. Scanning the Microwind folder I found several sample files. I loaded the 3-stage VCO circuit and noticed the incredible speed of the screen painting. I activated the Simulate command on the menu bar, and the SPICE plots appeared immediately. Not only was the code fast, it allowed me to simulate the circuit using a BSIM4 MOSFET model! The next feature I tried was the 2-dimensional viewer; this worked very well and could be used to display the patterned layering along any selected line. Then, I started the 3-dimensional simulator and watched in amazement as the program drew a three-dimensional perspective view of the chip as it was being fabricated.

I spent the next two weeks engrossed in studying the capabilities of Microwind (while attending to my normal duties, of course). I was so impressed with it that I decided to write this book. Professor Sicard, who had already started a book on submicron design, supported my proposal without hesitation. The two books evolved to be closely linked, with this as the basic introduction and his (with co-author Sonia Bendhia) being more detailed and advanced in both coverage and applications. Since there is minimal overlap, we think that you may find both volumes useful.

This has been the most enjoyable writing project I have ever undertaken. The Microwind and Dsch programs are extremely powerful tools, and exploring them was a lot of fun. I hope that you will find the programs as exciting as I do.

● ● ● ● ● ● ● ● ● ● ● ● ● ● ●●●

General Outline of the Book

The book has evolved to cover four main themes. The first group of chapters uses the Microwind program to illustrate how material layers are patterned to create a CMOS integrated circuit. The discussion covers the basics of the CMOS fabrication sequence and how it relates to using a layout editor. Design rules, parasitic resistance and capacitance, FETs, and general layout are examined.

The second theme is concerned with the electrical characteristics of MOSFETs as they relate to the layout. Simple analytic expressions are compared to SPICE models in Microwind.

CMOS logic circuits and chip design problems constitute the main portion of the book. The inverter is used to initiate the concepts of logic formation with MOSFETs, design of a basic library cell, and characterize the electrical performance. More complex cell functions are developed at the layout level.

● ● ● ● ● ● ● ● ● ● ● ● ● ●●●●

Acknowledgments

I am grateful to Professor Etienne Sicard for providing the basis for this book. Not only did he write the software, but he took the time to answer my never-ending questions and custom sculpt some of the features. While I'm sure my emails drove him crazy at times (the record was five in one day!), he always maintained his humor and wonderful outlook on life.

Bill Stenquist, my friend and colleague from Thomson, provided the motivation needed to undertake this project. Thank you Bill.

JOHN PAUL UYEMURA

There are so many that have made this book possible, I am not sure I can even begin to name them all, please excuse me if I have omitted anyone, it is not due to lack of appreciation. A special thanks to John R. Barry of Georgia Tech, for his help, and allowing me to impose upon him while in France. Thank you to all the staff of Thomson Nelson that help put this book into production, especially to Bill Stenquist, who *never broke his promise.*

Dr. Roger Webb, Dr. John Buck and Dr. David R. Hertling have my never-ending appreciation for all you have done for me and my family.

I also want to thank the many reviewers, especially Dr. Reginald Perry and Dr. Bruce Darling for taking the time to weed through the manuscript, and for the special care you took to help me produce a book that meets the higher standard that we all came to expect of the author. Thank you for helping me make sure I didn't let John down.

I want to especially thank Dr. Alper Eken for his dedication and review of the manuscript and for the perfect picture for the cover; and Dr. Amer Atrash, who always dropped what he was doing to assist me and was instrumental in the completion of the book. I thank Gale and Janet Sights, Donna Sylvan and Julie Young, for all their love and support. I was so very fortunate to have you there for me. Love and encouragement from my sister, Joyce McGough, grew my faith closer God, giving me what I needed the most.

The acknowledgements would not be complete without thanking Valerie and Christine, that have put up with me finishing this project. Their love sustained me and reminded me of what my life is all about. I will never forget the love that created them and for all the love and happiness they brought into our lives. *We love you both, so very much.*

MELBA UYEMURA

Foreword and Tribute

It is my great pleasure and pride to introduce John Uyemura's last book on integrated circuit design. Having graduated as John's first Ph.D. student in 1985 from Georgia Tech, I had the pleasure of interacting with him over the past two decades. I was part of the cadre of Georgia Tech EE students who loyally took *every one* of his new classes on integrated circuits and who flocked around his door during office hours for answers to additional questions and tidbits of his insight. His knowledge of the field was immense, but his charisma for exciting his students about the material was even greater. His courses became instantly popular and he soon began introducing new courses faster than many of his students could take them. Brand new classes would fill to capacity based solely upon the reputation that you would work like crazy but leave with a complete working knowledge and set of skills in the material that was covered. John was largely responsible for introducing modern IC device electronics into the Georgia Tech EE curriculum, which has flourished ever since.

Besides myself, several other of his Ph.D. students have entered the academic profession, doubtlessly because they too caught his contagious bug for teaching and scholarly work. Shortly after completing my Ph.D. degree, John began work on his first textbook, *Fundamentals of MOS Digital Integrated Circuits*.[1] This book, like the many others that would follow it, is a product of his classroom teaching and student projects. Because of the refinement that each set of notes went through in the classroom, John's textbooks quickly became famous for their readability, attention to detail, and up-to-date treatment of the subject matter. Most have been adopted at numerous universities, and have been translated into several different foreign languages. John's books have always had the feature that they could be easily read and understood by any student of the field, but even a seasoned expert could pick up a new thing or two from them. They have been both broadly accessible and technically complete treatments of the field, as were his classes. In many ways, this was his gift, being able to *explain almost anything to anybody*, and with a manner that always prompted a satisfied smile.

---

1. J. P. Uyemura, *Fundamentals of MOS Digital Integrated Circuits*. Reading, MA: Addison-Wesley, 1988.

John's relentless passion for teaching and textbook authoring are still clearly shown in this, his last book on IC design. In it, readers will find an amazingly clear and complete treatment of modern submicron CMOS VLSI design, with a concurrent explanation of and hands-on exercises with design automation software for automatic layout, design rule checking, circuit extraction, and simulation. Once again, John's book is right on target to where the field is going and to where students need a clear introduction. The book is skillfully written, broadly accessible, richly illustrated, and well synchronized to the Microwind and Dsch software that accompanies it. The organization and pacing shows that the material has been classroom tested and, perhaps more than any other feature, the book is simply a pleasure to read. This is a book written by an author who was at the peak of his talent, and who clearly received a lot of enjoyment in putting the project together. It is a perfect legacy to a gifted teacher, author, and friend, whose influence has changed so many lives and careers.

John was always happiest when he was teaching those around him. There was a certain point when he came into full form, crafting his lecture into a symphony of ideas, words, formulas, drawings, jokes, and analogies, which at once came together and opened up a new vista of understanding to those in his audience. And when he did that, he knew it, and you knew he knew it from that wonderful gleam in his eyes. He had you right where he wanted you sharing a moment of newfound understanding with him. John taught many of us the pleasures of teaching and of opening these new vistas to others. And so doing, he has also taught many of us what that gleam in his eyes was all about, as we now have it, too. Thank you, John.

ROBERT BRUCE DARLING
Seattle, Washington

About the Author

John Paul Uyemura
1952–2003

John P. Uyemura received his B.S., M.S., and Ph.D. degrees in Electrical Engineering and Computer Science from the University of California, Berkeley in 1974, 1977, and 1978 respectively.

Dr. Uyemura was known for his "pig problems" as the "Samurai Professor" after joining the School of Electrical Engineering at the Georgia Institute of Technology in 1978, where he taught for over 25 years. As a dynamic and demanding teacher, Dr. Uyemura's passion for teaching fueled the writing of over six textbooks, of which have become standards in their fields.

He pioneered academic programs in microchip and integrated circuit design, fiber optics, and VLSI design. He was also a leader in high-speed mixed signal CMOS circuit design. Dr. Uyemura consulted with IBM, AT&T, Furukawa Electric Corp., GEC Avionics, Amdahl Computers, and Bell South Corporation.

Dr. Uyemura was voted Georgia Tech's Outstanding Teacher of the Year and he was recognized for his role in co-developing Georgia Tech's first fiber optics instructional laboratory.

Dr. Uyemura is the author of *Fundamentals of MOS Digital Integrated Circuits*, (Addison-Wesley, 1988), *Circuit Design for CMOS VLSI*, (Kluwer Academic Publishers, 1992), *Physical Design of CMOS Integrated Circuits Using L-Edit*, (Brooks-Cole Publishing, 1994) *CMOS Logic Circuit Design*, (Kluwer Academic Publishers, 1999), *A First Course in Digital Systems Design*, (Brooks-Cole Publishing, 2000), *Introduction to VLSI Circuits and Systems*, (John Wiley & Sons, New York, 2002).

Installing the Microwind Software

Microwind is an integrated chip layout and simulation package that was written by Professor Etienne Sicard of INSA (*Institut National des Sciences Appliquées*)[1] in Toulouse, France. The software was written as an aid for learning submicron Complementary Metal-Oxide Semiconductor (CMOS) integrated circuit design, and it has many features that make it unique. Microwind combines various computer tools in a single package that allow the user to layout, check, and simulate a CMOS circuit interactively. It also has a compiler that can create the layout of a logic gate directly from a Boolean expression. When interfaced with its companion program, Dsch, one has an entire automated design environment on a desktop (PC). Logic schematics can be designed and tested in Dsch and then transferred to Microwind for compilation into a silicon CMOS circuit.

The compact discs (CD) provided inside the cover of this book has copies of both the Microwind and Dsch programs. The rest of the book uses Microwind to provide a "hands-on" approach to learning chip design. Dsch is only used in two of the later chapters in the context of automated design techniques.

This chapter will help you get the programs installed and running on your computer so that you can follow the discussion. Layout is an art, and practice is important. However, the treatment does not assume any previous experience in CMOS physical design.

● ● ● ● ● ● ● ● ● ● ● ● ● ● ● ●

1.1   Getting Started

Microwind was written for personal computers (PCs) that use Microsoft® Windows® as the operating system. It has been tested on Windows 98 and 2000 by the author, and others

---

1. Translation: *National Institute of Applied Sciences.*

have verified its operation on Windows 95 and NT/XP. The minimum requirements for installing and running Microwind from the CD are as follows:

- A CD-ROM reader
- A hard disk drive with at least 5 MB free
- A 128 MB system RAM is recommended
- A microprocessor with a clock frequency of 400 MHz or faster is recommended for the circuit simulations
- A mouse is recommended over other pointing devices
- A color monitor with VGA or better resolution
- A printer is needed to obtain hardcopies of your designs

The processor speed determines how fast the simulations and graphics operations work, but the program is designed to operate on a wide range of machines.

To install Microwind, insert the CD into the drive to initiate the autoboot process. If the program does not start, double-click on the My Computer icon on the desktop, select and double-click on the CD icon, then find the Setup icon and double-click on it. Alternately, you may follow the sequence:

Start
Run (type in x:setup where x is the designation of your CD drive)

This process will install the program on your hard disk and place a Microwind shortcut icon on desktop. Or simply create a new folder, name it Microwind, open up the CD, and copy all the files from the CD into the new Microwind folder.

● ● ● ● ● ● ● ● ● ● ● ● ● ● ⋯

1.2   Exploring Microwind

To start Microwind, open up the Microwind folder and then double-click on the desktop icon that is named Microwind2.exe. You may want to create a desktop shortcut as shown in the margin, to it so that you will have fast access to the program in the future.[2] This will launch the program and give the screen shown in Figure 1.1. This contains all of the primary menu commands and is the screen that we will always start from.

The work area on the screen has a black background with a set of visible grid points, but is shown as a white area in the picture for clarity.[3] **Menu** buttons are arranged along the top of the work space, with the normal menu words above them. When we refer to menu commands in the text, we will use a boldface sans serif font: **Menu Command**

---

2. To create a shortcut on your desktop, point to the icon and right-click to produce the Windows menu. Select Create Shortcut from the menu. This will create an identical icon with a "2" appended to the name. Then, drag the new icon to the desktop.

3. It is possible to change the background to white, but we will keep it as is for now.

Figure 1.1:
Microwind launch
screen

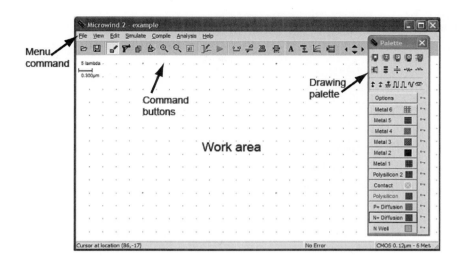

so that you will be alerted to look for them on the screen. The window on the right side of the screen is called the drawing Palette menu. As we will see, the Palette menu is our entry point to designing circuits in silicon. Each of the colors represent a specific material used in the fabrication of a chip. Designing a chip at the silicon level requires that we specify how each material is shaped and stacked with the others to form transistors and wires. The Palette menu allows us to specify each material as we design the chip and has several useful macros that simplify the task.

Let us specify a few notational conventions that will make it easier to explore the program. Microwind is a mouse-driven program, and most of the commands require that the user use the left mouse button. The push-and-release action will be called a "click." We will use the expressions "single-click" or "click" for one-click commands, and "double-click" for two-click commands, respectively, with the left mouse button implied. Alternate terminology is "left-click" or "right-click" for single-click commands; this allows us to distinguish between the buttons if necessary.

If an action requires simply pushing an on-screen "button" icon, we will often provide a screen shot the first time it is used to help you learn the program. For example, the **Copy** button is shown in the margin. Many of the most important Microwind commands can be accessed using buttons, and the icons have been selected to make them easy to remember. You can always determine what a button does by pointing to it with the mouse cursor, as this will cause a descriptive dialog bubble to appear on your screen.

Copy button

Most drawing actions and some commands require you to push the mouse button and hold it down until the desired affect is achieved. We will refer to this action as "push and hold," and it is always done with the left button. When it is necessary to change the position of the mouse pointer, we will say to "drag the mouse."

Menu commands generally require you to make a sequence of selections starting with a main heading, such as **View** or **Edit**. To sequence through this type of operation, point to the Menu option until it is highlighted, and then left-click to bring down a submenu.

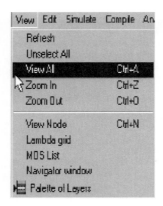

Figure 1.2:
Menu example

Select the command on the submenu by pointing to it until it is highlighted, and then left-clicking the mouse button. In the text, we will specify this type of a sequence using a special type format:

First menu command ⇒
   Second menu command

The arrow (⇒) indicates that you should move the next submenu window to make a selection. Screen images will often be inserted in the text to help you sequence through nested submenu when they occur. An example is shown in Figure 1.2 for the sequence:

View ⇒
   View All

Clicking on the main menu command **View** opens the window containing the submenu options. To execute the command **View All**, single-click on the appropriate line.

Some Microwind commands can be also be entered from the keyboard. In the text, keys will be specified using boldface square brackets, e.g., **[Z]** implies the "z" key (lower case). If a command requires simultaneous keystrokes, they will be listed together. As an example, **[Ctrl] [Z]** means that you should hold down both the **[Ctrl]** and **[Z]** keys at the same time; the simplest approach is to hold down **[Ctrl]** and then strike the **[Z]** key. The sub-menus list the keystroke alternatives when they exist. For example, the highlighted line in Figure 1.2 indicates that the command **View All** can be executed from the keyboard using **[Ctrl][A]** as written on the right.

The simplest approach to learning the Microwind program is to work with it. Some of the main features are illustrated in the following sections, and you should work through each to get the "feel" of how the software works. Don't worry about understanding the details for now. They will become clear as we progress through our adventure in chip design.

We should comment on the differences between the screen dumps printed in the text versus the images that appear on your computer screen. When you work through the examples, you will see that Microwind uses full-color displays. This book has been

Figure 1.3:
Opening a file

printed using greyscale shading to reduce costs, since providing full-color printouts would have quadrupled the price. When we refer to colors in the text, they will be the ones that you see on your screen. Also, several background screens in Microwind are black; in most cases, these have been changed to white for easier reading.

1.2.1   Studying a Design

Let us begin our study by examining a pre-designed circuit file in the Microwind folder. Assuming that you have launched Microwind, execute the following command sequence:

File ⇒
   Open

The **File** submenu is shown in Figure 1.3. When executed, this command opens a window that lists all of the available files. The **File** button shown to the left may also be used to execute the command sequence.

File button

   The collection of files supplied in the Microwind folder is quite large, and you must use the arrow button on the right side below the file list to see all of the entries. For this example, we will look at a file named **Inv3.MSK**. You will notice that the list is in alphabetical order, so that we need to move to the right. Scrolling gives the view shown in Figure 1.4. Clicking on the icon opens the file and gives us the image shown in Figure 1.5. This is an example of a **chip layout**. Each color represents a different layer of electrically conducting material, and the patterns combine to form electronic switches and to show how they are wired together. The drawing on your screen is in color, but the one shown in the drawing is a monochrome version that uses different cross-hatching schemes instead of different colors. If you would like to see the monochrome view, push the **[F5]** function button. The change can be made using the menus from the sequence:

File ⇒
   Colors ⇒
      Switch to monochrome

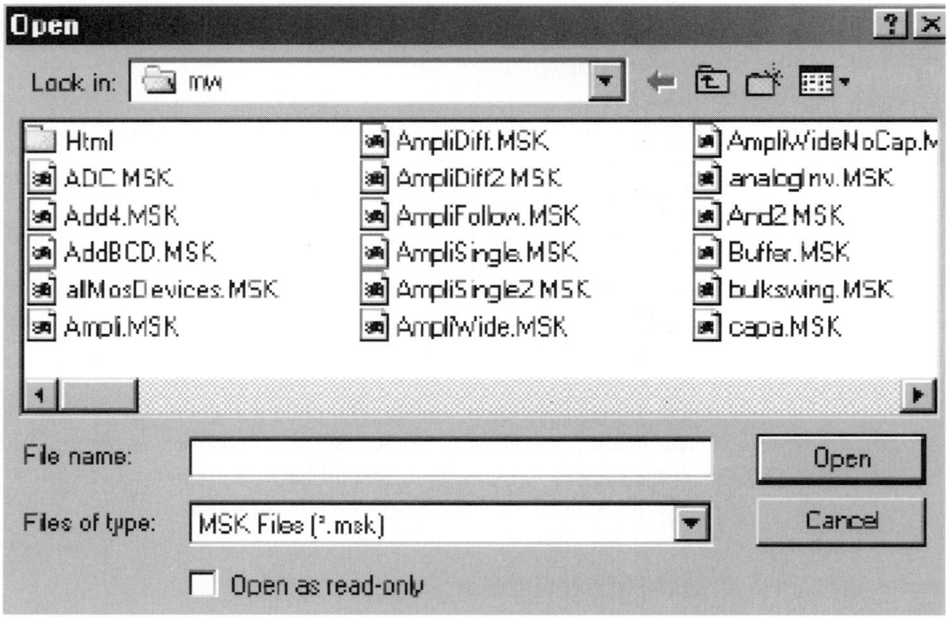

Figure 1.4:
Finding the
Inv3.MSK file

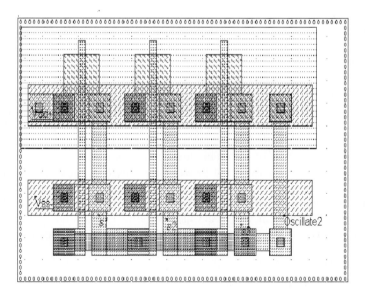

To return to the color screens, either push [**F5**] again or go through the command sequence again and switch back to color.

A significant amount of chip design deals with creating layout drawings like that shown in Figure 1.5. A layout plot represents an electronic circuit as it would be built in

Figure 1.5:
Layout for the
Inv3.MSK file

silicon. The plot is designed on a computer, and the corresponding data files provide the chip manufacturing plant with all of the information needed to actually produce the device. The dimensions of each patterned regions are critical to the electronic operation, and changing the size of even a single rectangle may introduce significant consequences. We will spend a lot of time learning how to design layout plots like these. Once you learn the basics, you will be able to read the layout like a schematic diagram and visualize the transistors and wires without much effort.

The chip layout represents the chip as viewed from the top looking down. Chip design is always done from this perspective, but sometimes it helps to see 2- and 3-dimensional views of the structure. Microwind has built-in routines to provide both in the submenu under the **Simulate** command.

Let us look first at a 2-dimensional cross-section. This will show us what the chip would look like if we could select a line on the top-view layout and slice the structure downward. Starting from the layout screen, we use the command sequence:

Simulate $\Rightarrow$

Process section in 2D

The menu display for this type of sequence can be seen in the earlier example of Figure 1.2. Shortcuts for executing this command are to use the keyboard sequence **[Ctrl][2]** or the button shown on the left; the button icon is very easy to remember as it is of a saw cutting a piece of wood. By itself, the command does not induce any action *until* you specify where you want the cut to be. Place the mouse pointer on the left side of the layout, then push and hold the left button to specify one side of the line. Drag the mouse pointer to the right while holding the left button down; you will see a thin line that follows your motion. The line indicates the path where the cross-section will be taken. Releasing the button specifies the other end of the line and results in a screen that looks something like that shown in Figure 1.6. Although the general features will be similar, your screen view will be different because your line is probably different than what was selected for the text. The

2-D button

Figure 1.6:
2-dimensional view
of the layout plot

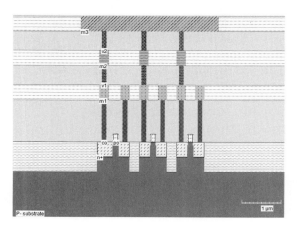

important features are the different color-coded layers and the sizes of the individual features shown in the 2-dimensional view.

A cross-sectional view of an integrated circuit shows that it is made up of many different layers, each represented by a different color. For example, the blue sections represent different metal layers, while the red sections are a special type of silicon called **polycrystalline silicon** (just **poly** for short). The layers are listed in the window on the right side of the screen. Note that there are several layers listed that do not appear on this particular plot. The height of each colored section is determined by the thickness of the layer it represents, while the width of a feature varies with the layout. To exit the simulation, push the **OK** button at the bottom-left side of the screen:

This places you back in the layout plot, which is always the starting screen for other actions. You may see other cross-sectional views by selecting other lines. The 2-dimensional viewer will be active, so long as the on-screen button is pushed; a pushed button is outlined.

Layer List		
Layer	Thick(μm)	Height
metal6	0.70	6.60
via5	0.50	6.10
metal5	0.70	5.40
via4	0.50	4.70
metal4	0.50	4.20
via3	0.50	3.70
metal3	0.50	3.20
via2	0.50	2.70
metal2	0.50	2.20
via	0.50	1.70
metal	0.50	1.20
poly	0.20	0.01
poly2	0.20	0.22

Layer listing

A striking and unique feature of Microwind is the ability to create a 3-dimensional view of the circuit in a simple manner. The command sequence is

3-D button

Simulate ⇒
Process steps in 3D

The keyboard equivalent is **[Ctrl][2],** or you can use the button shown at the left. Executing this command opens a window that shows a 3-dimensional perspective of the silicon based region. As we saw earlier, a silicon integrated circuit consists of several layers of materials, with each layer patterned to create the electronic network. The layers will have more meaning once you see how they really look.

There are different options for creating the 3-dimensional plot. The simplest is to let Microwind build the drawing for you by clicking on the **Auto** button:

This starts an animation sequence that shows each layer being added to the structure to build the chip. Alternately, you can use the **Next Step** and **Previous Step** buttons to change the view manually.

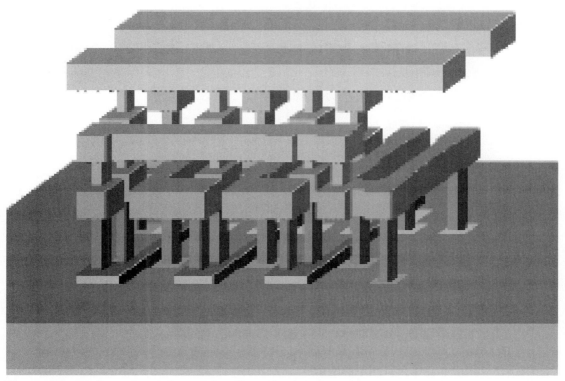

Figure 1.7:   3-dimensional view of the silicon circuit

This simulation is based on the actual fabrication sequence, and helps you visualize the characteristics of the chip structures that are important in advanced design. The final result will be something like that shown in Figure 1.7. Your screen may be different depending upon the resolution of your monitor. The arrow buttons can be used to pan the drawing up or down as needed. After you have finished looking at this view, click the **OK** button to return to the layout screen.

The geometrical patterns that you see in the top view (layout) screen summarize a large part of what is required in chip design. The dimensions of each pattern on the layout are determined and carefully drawn by the designer to insure that it is possible to actually build the structure within the limits of the fabrication process. The patterns portrayed in the layout represent the components of an electronic network. In chip design, the electrical characteristics of each component are determined by the shape and dimensions of the patterned regions. The overall network response parameters, such as the switching time or maximum frequency, thus are related intimately to the layout. If we change the shape, size, or position of a region, the electrical response will also change.

Figure 1.8:
Simulate menu

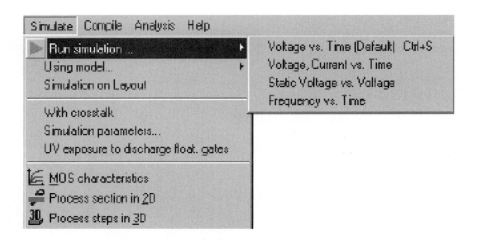

Microwind has a built-in circuit simulator that can be accessed with the command sequence:

Simulate ⇒
Run Simulation

Run Simulation
button

and then select the type of simulation. The windows are shown in Figure 1.8. A simpler approach is to use the **Run Simulation** button (shown to the left). The default simulation choice is to have Microwind plot voltages as a function of time. This can be invoked by the keyboard command **[Ctrl][S]** directly.

For this example, select the default simulation on the first line. This gives a set of plots that show the voltage waveforms as functions of time at various nodes in the circuit. The **Inv3.MSK** file produces the screen that is shown in Figure 1.9.[4] You can use the cursor to measure voltage and time differences directly on the plots. For example, point to a spot on a waveform, click and hold the left button, and then drag it to the right. When you release the button, a timing ruler will appear and tell you the time interval spanned by the line you drew. The Control window on the left allows different aspects of the simulation to be displayed.

Other options in the Simulate window produce different plots. They can be selected using the tabs at the bottom of the screen. Some may have meaning for this circuit, while others may produce blank screens. The outputs that are available depend on the sources and conditions for each circuit. Some simulations may alter other calculations; pushing **Reset** starts the run over again.

This string of exercises illustrated some of the most important features of Microwind as an analysis engine. However, we have not even begun to show you the overall power of the program. It also serves as a layout editor, a component design tool, and a process and

---

4. If you do not get this screen, some information may have changed in the layout screen when you were exploring the 2-dimensional and 3-dimensional options. For now, just **Open** the file again and run the simulation immediately after it loads. We will learn how to adjust the SPICE simulations later.

Figure 1.9:
Circuit simulation
waveforms

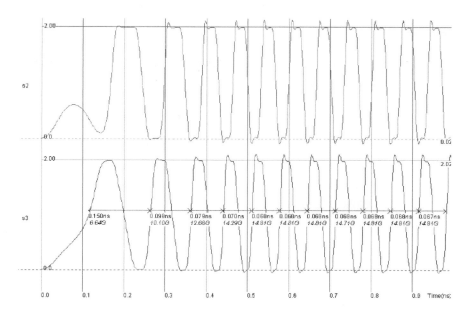

device characterization aid. The program can be used to design chips that range in complexity from simple gates to complex microprocessors. Your work can be saved internally, or exported to standard formats for others to use. **Print** buttons on all major screens allowed you to document your work on hardcopy.

1.2.2   Looking for Help

Microwind has many commands to help you design chip layouts and simulate them. We will present most of them in this book and provide examples of how they are used. However, the program itself has an online reference manual that provides basic information on all important features. It can be accessed using the sequence:

> **Help** $\Rightarrow$
>> **Reference Manual**

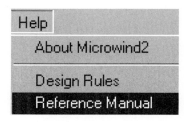

Figure 1.10:
Quit command

The online reference manual is organized to help you find the meaning and usage of each command.

1.2.3 Leaving Microwind

When you are finished working in the Microwind environment, you may quit the program using either the **File** menu sequence

> **File** ⇒
>> **Leave Microwind2**

as shown in Figure 1.10, or the keyboard shortcut **[Ctrl][Q]**. Depending upon what commands have been executed, you will get one or two dialog windows. The first will ask you if you want to save any changes to the file, while the second will confirm that you want to leave the program. The appearance of the dialog windows are provided in Figure 1.11. Clicking Yes confirms your decision. It is recommend that you not save changes that you make to any files that are supplied with this Microwind program, as several will be accessed later in the book and need to be kept intact. If you do alter files, you can always restore them from the original Microwind program disk.

Figure 1.11:
Dialog boxes for
leaving Microwind

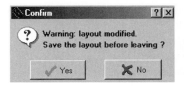 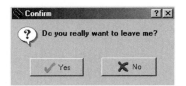

(a) Save layout dialog   (b) Final confirmation

1.2.4   **Simulation Limits**

Microwind allows you to design and create large CMOS circuits. The present version has been compiled with the following limits:

- Maximum number of text labels: 500
- Maximum number of layout boxes: 20,000
- Maximum number of electrical nodes: 500
- Maximum number of transistors: 500

If you have only a small amount of system memory installed on your computer, it is possible that you may experience some size limitations. However, all circuits in this book were designed and simulated on PCs with 128 MB without any problems. In addition, all of the design examples in the Microwind folder have been tested.

● ● ● ● ● ● ● ● ● ● ● ● ● ● ● ● ●
1.3   Installing Dsch

The CD also contains another program called Dsch. This is an excellent logic simulator that has advanced modeling capabilities. It also forms the front end of an automated logic compiler that works with Microwind to turn a logic schematic into a CMOS integrated circuit.

This book centers around the physical design of CMOS integrated circuits which makes Microwind the central program. Dsch is introduced in Chapters 15 and 16 in the context of design automation. The fundamentals of using the Dsch program are covered in Chapter 15, along with the use of the silicon compiler. Chapter 16 presents examples of the automated-design process.

If your interest is in quick CMOS layouts that are, for the most part, completely automated, then it is possible to skip directly to Chapter 15 and start the basics there. You probably will need to step back and review the other chapters to fully appreciate the power and limitations of the automated-design flow. Access to Dsch is required to follow the discussions in Chapters 15 and 16, but it is not needed for any of the earlier chapters. Since there will be those of you that may want to learn how to use the program earlier, we include the installation instructions here.

To install Dsch, use the following procedure.

- Place the CD in the drive
- Double-click the My Computer icon on the desktop
- Open the CD drive by double-clicking on it
- Point to the Dsch folder, click and hold the left button, then drag it to the desktop. Release the button to copy it to your computer

You can move the folder off of the desktop using standard Windows techniques, if you want. Once it is in place, make a shortcut using the following procedure.

- Double-click on the folder to open it. Then find the Dsch icon, make a shortcut by pointing to it, and using the right mouse button to find the command, then drag the shortcut icon to the desktop
- Close the Dsch folder

To start Dsch, simply double-click on the Dsch shortcut icon.

1.4 Plan of the Book

This book has been written to provide you with a hands-on approach to learning the basics of CMOS layout and circuit simulation using Microwind. The techniques are very general and can be used with any chip design toolset. Only a minimal background in digital logic and basic electronics is assumed. No specific background in integrated circuit design is needed to follow the discussion.

Integrated circuit layers are presented in Chapter 2. These general concepts form the basis for a step-by-step examination of the CMOS fabrication sequence in Chapter 3. The fundamentals of layout are covered in Chapters 4 and 5 and complete the general introduction to layers and patterns.

The electronics aspects of chip design start with a discussion of MOSFETs in Chapter 6. This chapter presents the theory of operation from an analytical and physics-oriented viewpoint. This prepares the reader for the discussion of SPICE modeling in Chapter 7. Once devices have been covered, CMOS logic circuits are introduced in Chapters 8 through 11. Interconnects are added in Chapter 12, and the basics of silicon-on-insulator (SOI) technology are presented in Chapter 13 using the Microwind model.

Chapters 15 and 16 have a somewhat different theme. Design automation is introduced with the Dsch program where logic circuits are designed and then translated into the Verilog hardware description language (VHDL). The code is then read by Microwind, which synthesizes the CMOS layout for the circuits. These two chapters illustrate the Dsch–Microwind design flow with many examples and also introduce general concepts from the field.

The last two chapters present layout basics for analog circuits. Chapter 17 discusses integrated capacitors and inductors in a CMOS technology. Layout considerations for analog circuits are introduced in Chapter 18.

 Microwind is referenced throughout the book, and step-by-step instructions are given the first time a command or technique is introduced. Some of the more important examples are denoted by the icon shown here. There are many layout files provided in the Microwind folder, but only a few actually are discussed in this book. The reader is encouraged to open them up and study them. Also, Microwind creator, Etienne Sicard, and his colleague, Sonia Bendhia, are co-authors of a companion book on submicron CMOS design. Many of the examples are analyzed in detail in their book.

● ● ● ● ● ● ● ● ● ● ● ● ● ● ⋯⋯

1.5   Some Important Details

There are a few items that are worth mentioning up front.

1.5.1   Copyright

The programs on the CD are copyrighted by INSA and Etienne Sicard. Professor Sicard has agreed to allow free usage and distribution of the program, but it is against international copyright law to use any portion of it in a commercial product or venture without the written consent of the copyright holder.

1.5.2   Versions and Updates

The programs provided on this CD are the versions available at the time this book was published. Since the program is under continuous development, there may be some differences in the printed screen dumps and what you see on your computer. These are usually minor, and new commands can be understood by just playing with them. No major changes in the overall characteristics or operation are anticipated in the near future.

Professor Sicard maintains a website where you can download the latest releases. The URL (and a nice photograph!) can be found by executing the

Help ⇒
About Microwind

command. Updates of this book are planned if any major changes in the program operations or capability occur.

1.5.3   Additional Help?

The online reference manual should answer most questions on commands and usage. In addition, a summary of the menus are given in Appendix A of this book.

Experience shows that most questions about commands can be resolved by simply trying them. Save your work frequently. If some operation fails, close the file, reopen a saved version, then try it again.

Since the program is an academic endeavor, not a commercial product, it is not possible to provide individualized assistance. While the author is interested in your feedback, time and resource constraints limit his ability to respond to questions posed in e-mails or letters. All of the questions that have come up while testing the program and writing this book have been addressed. The publisher cannot answer any technical questions on the use or operation of the program.

● ● ● ● ● ● ● ● ● ● ● ● ● ● ● ●

1.6 References

This book deals only with the physical design aspects of VLSI. While this is related directly to circuits, electronics, logic, and other areas, they are introduced and referenced somewhat sporadically. Although the book can be read independently, it is anticipated that most interested persons will either be taking, or have completed, a course in VLSI design.

The books listed below deal with general aspects of digital VLSI and CMOS circuits that complement the treatment here.

[1.1] Smith, M. J. S., *Application-Specific Integrated Circuits*. Reading, MA: Addison-Wesley, 1997.

[1.2] Uyemura, J. P., *A First Course in Digital Systems Design*. Brooks/Cole, 2000.

[1.3] Uyemura, J. P., *Introduction to VLSI Circuits and Systems*. New York: John Wiley & Sons, 2002.

[1.4] Uyemura, J. P., *CMOS Logic Circuit Design*. Norwell, MA: Kluwer Academic Publishers, 1999.

[1.5] Wolf, W. H., *Modern VLSI Design. Third Edition* Upper Saddle River, NJ: Prentice Hall, 2002.

Views of a Chip—Layers and Patterns

Chip design is a fascinating field to study and work in. Using a sophisticated set of computer-aided design (CAD) tools, one can envision a universe that is completely different from the one we live in. In this microscopic world, electrons travel through reconfigurable mazes that perform logic and are created by the chip designer. In this chapter, we will introduce you to the world of a chip designer as seen at the physical silicon level.

● ● ● ● ● ● ● ● ● ● ● ● ● ● ● · · ·

2.1 The Design Hierarchy

Integrated circuits (ICs) are microscopic electronic networks that are created in a special type of material called a **semiconductor**. Silicon is a semiconductor and is used as the base material for the vast majority of modern electronic systems. As integrated circuits became more commonplace, writers started using the phrase **computer chip** to describe the devices used to create the computer revolution, and the term has evolved to be accepted by all. The descriptive word *chip* arises from the physical appearance of the prototype silicon integrated circuit with both digital and analog circuit functions. Figure 2.1 is a photograph of a silicon integrated circuit that has side dimensions of 2.7 mm × 2.6 mm.[1] The actual size of the circuit is shown here:

This example shows that it really does look like a "chip" of some sort!

Although you may think that computer chips have been around forever, integrated circuits were not invented until 1956 by Jack Kilby of Texas Instruments Corporation. The

---

1. This chip was designed in a joint IDT-Georgia Tech project by Dr. Brian Butka (IDT), Dr. Y. Alper Eken, Dr. Paul Murtagh (IDT), and Dr. Bortecene Terlemez.

Figure 2.1:
Photograph of an
experimental CMOS
telecommunication
chip

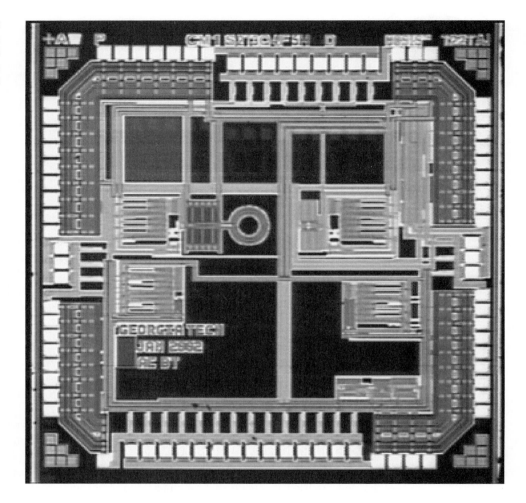

significance of this invention is reflected by the fact that Kilby shared the 2000 Nobel Prize in physics for his work. Integrated circuits gained steady acceptance and usage as our knowledge and design skills increased and the cost per chip decreased. The development of the personal computer (PC) in the 1980s fueled the "computer revolution," and the unparalleled explosive growth of modern computer systems and networks has provided the driving force to the present day.

Integrated circuits are quite complex, and a complete understanding of every aspect of chip design and fabrication requires several years of study and practical experience. This task is made easier by breaking the problem into **design hierarchies** where the problem is viewed at several different levels. Each level concentrates on a specific viewpoint, such as logic design, that can be linked to other levels. Some levels are abstract, while others can

Figure 2.2:
Structure of the
design hierarchy

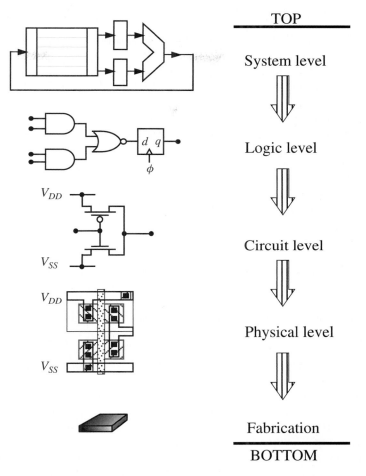

TOP
_____

System level

⇓

Logic level

⇓

Circuit level

⇓

Physical level

⇓

Fabrication
_____
BOTTOM

be visualized very easily. Figure 2.2 shows the basic breakdown of the hierarchy for a digital integrated circuit. Starting at the top of the drawing, we see the highest level is:

- **System Design**—Where the main operations of the chip are determined. Block diagrams are used to illustrate the main sections that make up the system. Only the input/output characteristics are important, and there are no details about what actually is inside each block.

The next level down is that of

- **Logic Design**—This is where the engineering teams design the logic networks that are required inside each block to obtain the input/output characteristics used at the system design level. The output is generally in the form of a netlist, which is just a description of the logic gates and wiring needed to implement the design.

Once this is accomplished, we move to the next lower level of

- **Circuit Design**—Where the logic network is transformed into an electronic network using transistors as switching devices. Digital variables are represented by voltage levels that change in time. Transistors allow the designer to create logic circuits that steer signals into different paths using switching mechanisms.

Once the circuit design is completed, it is transformed into a design file for a silicon integrated circuit in the next level that is designated as

- **Physical Design**—In this step, the electronic circuits are transformed into on-screen colored geometrical patterns using computer graphics and analysis tools. Each color or shading represents a material, such as a metal, and the patterns indicate how to form 3-dimensional transistor structures and wire them together.

The final step is that of

- **Chip Fabrication**—Where the physical design is transformed into a finished silicon chip that can be put into a package for eventual wiring into a product.

This sequence illustrates the **top-down** approach to chip design that is used for modern digital-chip designs. Project engineering is performed by groups at every level, and information exchanged among the various design levels is crucial. Many sub-tasks will overlap, but the hierarchy usually remains visible. The size of the project group varies with the complexity of the chip. While a basic, functional chip may be completed using only four or five main designers, a state-of-the-art microprocessor chip can require several hundred engineers, technicians, and support personnel for successful completion.

This book does not attempt to cover the entire design hierarchy, but concentrates on the study of physical design as a central theme. However, the close-knit characteristic of the design hierarchy leads us to other levels in a natural manner. This is because physical design is based on the characteristics of the fabrication sequence and forms the link to circuit and logic design. The Microwind software provides this link in an integrated manner and allows us to see the relationship among these three levels in a seamless manner.

● ● ● ● ● ● ● ● ● ● ● ● ● ● ●
2.2 Integrated Circuit Layers

In its simplest form, an integrated circuit can be viewed as a set of *patterned material layers*. If you worked through the examples in Chapter 1, then this statement already has some meaning. It does, however, deserve a more detailed discussion before we go into the intricacies of layout design.

A chip is the physical implementation of a signal control or processing unit. The signals change in time and are represented by voltages $V(t)$ at different points on the network.

Figure 2.3:

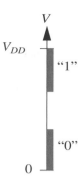

An example is shown in Figure 2.3. If we assume that the voltage can vary from 0 V to a peak value *of* V_{DD}, then we can associate binary equivalents with different ranges. In CMOS, the standard designation for the power supply voltages applied to a chip are

V_{DD} = positive power supply voltage

V_{SS} = ground or 0 V

The actual value of V_{DD} varies with the fabrication process. Some common values are V_{DD} = 5 V, 3.3 V, 2.5 V, and 1.2 V, but others are also found in practice.

The **positive logic** convention assigns a binary "0" to the low voltages, while a binary "1" is represented by high voltages. The intermediate values have no meaning in the Boolean world. Theoretical considerations provide the large-scale model of the system, but the final realization is a network that operates on the basic principles of currents and voltages.

To a chip designer, every material is characterized by how well it conducts electrical current. Metals are classified as **conductors**, meaning that they conduct electricity very well. Aluminum (Al) and copper (Cu) are the most common conductors used in CMOS integrated circuits. An **insulator** is exactly the opposite: it resists the flow of electrical current. If you examine a section of "wire," such as the type that is used to connect speakers to an amplifier, you will see that the electrical current flows on a metal strand that is encased by a plastic insulating jacket. Figure 2.4 shows this structure. The insulating cover insures that unwanted conduction paths will not be set up between two lines that happen to touch.

Figure 2.4:
Structure of a
speaker wire

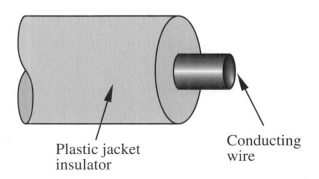

Plastic jacket
insulator

Conducting
wire

Figure 2.5:
Metal and insulator
layers of a chip

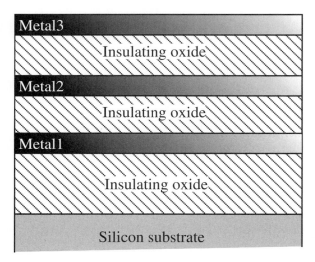

This simple idea provides the basis for integrated circuit design. Starting with a flat-surfaced area of silicon, we use various techniques to create an alternating stack of thin metal and insulating layers, as portrayed in Figure 2.5. The silicon itself provides the base for the structure, and is called the **substrate**; it is usually about 0.3–0.5 mm thick. The substrate is coated with a layer of insulating material such as **silicon dioxide (SiO_2)**, otherwise known as quartz glass. In our drawings, the insulating layers are generally referred to as oxides or glass, regardless of their actual composition.

The metal layer closest to the silicon is called Metal1, since it is the first to be deposited on the structure. After another layer of oxide is added, the second layer of metal (Metal2) is deposited on top. This is followed by another layer of insulating oxide, which is followed by another layer of metal shown as Metal3 in the drawing. This alternate oxide–metal layering scheme can be continued, and modern chips have eight or more layers of conducting metal.

Each metal layer is electrically isolated from all others by the insulating oxides. If we go one step further and define distinct patterns of **conduction lines** on the individual metal layers, then we can independently steer electrical signals on each level. This is analogous to creating small rectangular cross-sectioned wires, and is illustrated in Figure 2.6. The view in Figure 2.6(a) shows a 3-dimensional perspective for distinct patterns on the Metal1 and Metal2 layers. When these are stacked on the silicon, they give us the top view shown in Figure 2.6(b). This drawing uses different shadings/patterns to distinguish between the two metal layers, and we don't show the insulating oxide (since it is glass anyway). The significance of this structure is that points A and B are electrically connected through the Metal1 pattern, but are isolated from Metal2 points C and D. If we want a Metal1 line to contact a Metal2 line, then we can create a hole in the insulating oxide layer and fill it with metal that touches both metal layers. This is generally called a **contact** or a **via** and allows us to send signals on any desired layer.

Figure 2.6:
Patterned metal
layers in an
integrated circuit

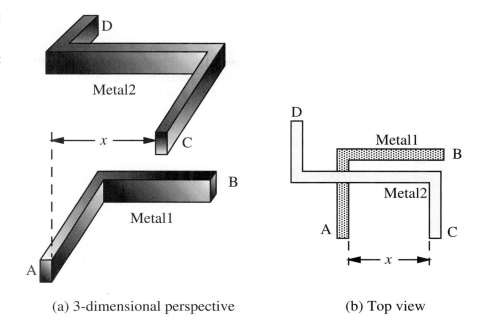

(a) 3-dimensional perspective   (b) Top view

A silicon integrated circuit uses patterned metal lines to provide conduction paths between points in the electronic circuit. Having access to several metal layers eases the problem of connecting a complex circuit together. Switching functions are implemented using controlled switching devices called **transistors** that are built into the silicon level by creating patterns of regions with different conduction characteristics. Control is achieved using a conducting layer above the silicon level. As you can see, a silicon chip really is just a set of patterned material layers.

We are finally in a position to define what this book is about. A digital integrated circuit is a logical switching network that consists of transistors wired together in a very specific way. The physical design of a silicon chip centers on creating patterns on each layer of an integrated circuit such that the resulting 3-dimensional structures form transistors and the wiring that connects them together. The shapes and sizes of each pattern are important to determining the electrical performance of the chip. This includes parameters such as the switching times, the power consumption, and the total area of the chip. Launching Microwind places you in a **layout editor**. This is a CAD[2] tool that is used to design the patterned layers of the integrated circuit. It gives you the ability to do physical design in a graphical environment. The electrical characteristics are obtained by using a circuit simulator that is built into the Microwind package. Our treatment will teach you the fundamentals of physical design as it is constrained by the fabrication process. We will also examine how our design translates into electrical characteristics using a SPICE[3] circuit simulator.

---

2.   CAD is an acronym for **C**omputer-**A**ided **D**esign.

3.   SPICE is an acronym for **S**imulation **P**rogram with **I**ntegrated **C**ircuit **E**mphasis.

The treatment will include both text discussions and hands-on PC examples that have been designed to reinforce your understanding of the material. Physical design is best learned by hands-on experience, so you should make an effort to follow the examples on your computer. Microwind is unique in that it integrates many useful engineering and pedagogical features in a seamless manner. You will probably find it extremely useful for studying integrated electronics far beyond what is covered here.

2.3 Photolithography and Pattern Transfer

A key concept in chip design revolves around the ability to create patterned material layers to steer electrical signals on the chip. This is accomplished using a technique called **photolithography** that allows us to translate on-screen computer drawings to a physical structure that replicates the patterns defined by our CAD package. The resolution of the photolithographic process determines the minimum feature size that can be fabricated on the chip, and is a limiting factor in the design of high-density integrated circuits. The sequence that is used to create the pattern can be summarized as

- Design the pattern on a computer
- Create a mask or reticle
- Print the pattern onto the surface of the chip
- Use the printed region to define the material pattern

Since this may seem a bit mysterious, let us delve deeper into this important aspect of chip design and manufacturing. Constraints introduced by photolithography have a great effect on how we design circuit layouts, making this a priority in our study.

The starting point is the computer data base that defines each polygon and its location on a layer. This information is translated into a standardized format that identifies each geometrical object by its vertices. Figure 2.7 illustrates the conversion to a Caltech intermediate form (CIF) file format. The file is then used to create a **mask** or **reticle** that is used to optically transfer the original pattern to the chip surface. The structure of a basic

Figure 2.7:
Translation of a pattern to a CIF file

```
L 13;
P 780,3360 960,3360 960,4020 780,4020;
P 10380,-420 10560,-420 10560,840 10380,840;
P 9600,-780 9960,-780 9960,-420 9600,-420;
P 2640,2040 2760,2040 2760,3360 2640,3360;
P 10380,3360 10560,3360 10560,4020 10380,4020;
P 780,1560 960,1560 960,2040 780,2040;
P 9600,4020 9960,4020 9960,4380 9600,4380;
```

(a) On-screen plot   (b) CIF file description

Figure 2.8:
Structure of a
basic reticle

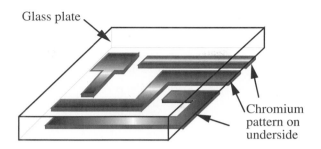

Glass plate

Chromium
pattern on
underside

reticle is shown in Figure 2.8. It consists of a high-quality piece of glass with a chromium metal replica of the pattern on one side. The dimensions of the reticle pattern are larger than the final values that will be transferred to the chip. This allows finer precision for small-geometry devices. For use in photolithography, the main idea is that the reticle is optically transparent except in those regions where the chromium metal regions exist; there, the reticle is opaque and any incident light is reflected by the metal. In the chip manufacturing process, each layer is patterned separately from every other layer. Thus, one mask/reticle is required for each layer. Mask sets for an entire chip are extremely expensive to fabricate, with costs exceeding $1,000,000 per set not uncommon in a state-of-the-art design. One reason for the high cost is that the features on a modern process are so small, a simple transmission/reflection technique is not sufficient to define the details. Many advanced optical methods based on diffraction and other effects must be used to create a good reticle. This complicates the process considerably, but it is a mandatory part of the chip-making procedure.

Transferring the reticle pattern to the surface of an integrated circuit takes place in the exposure sequence. The process starts by coating the surface of the chip with a light-sensitive organic polymer (i.e., a plastic) called **photoresist** (or just **resist**) which acts similarly to ordinary photographic film. When illuminated by an ultraviolet (UV) light source, regions of the photoresist that absorb optical energy experience an internal chemical change; if we can shield regions from the light source, then they are not affected. Developing the photoresist in a special solution yields a layer where some regions are hardened, while other regions can be washed away with a solvent. There are two "types" of photoresist, positive and negative. In positive resists, the illuminated regions are soluble after exposure to light, while unexposed regions develop into hardened material. Negative resists are exactly the opposite: the regions that are exposed to light develop into hardened sections. The clear and opaque regions of the reticle must be specified in the layout and transferred to the reticle in the mask-making process.

Consider the sequence described by the cross-sectional views in Figure 2.9. The starting point in this sequence is a substrate with an oxide coating and a metal layer on top of the oxide. Our objective is to transfer a pattern to the metal layer. First, photoresist is applied by spin-coating the surface of the chip with a liquid solution, and then baking it for a few minutes to dehydrate the layer. This leaves a photoresist coating, as in Figure 2.9(a). Next, the reticle is placed between a UV light source and the chip so that it casts a shadow onto the surface of the photoresist; Figure 2.9(b) shows this step. Optical imaging components (not shown) are used to sharpen the edges of the shadow to insure a high-quality

Figure 2.9:
Resist application
and exposure

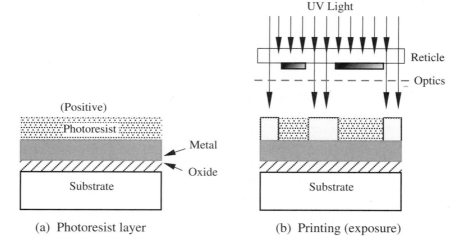

(a) Photoresist layer   (b) Printing (exposure)

print. This is called the **exposure step,** and only requires a few seconds to complete. After
the resist is exposed, it is developed and rinsed in a solvent; this leaves a hardened layer of
photoresist that has the same pattern as the original screen image.

The final structure after the developing procedure is shown in Figure 2.10. The important aspect of the drawing is the fact that the resist layer now contains the pattern information that was designed into the layout. Hardened photoresist itself is used as a masking
layer in the final step of the lithographic-transfer process. It acts as an efficient barrier to
chemical or gaseous agents that can be used to etch away the metal regions. This is called
the **etching process** and is shown in Figure 2.11. Modern IC fabrication employs **Reactive
Ion Etching (RIE)** in which ionized atoms of an inert gas such as argon (Ar) are mixed
with etch-assisting chemicals. The mixture is then excited with a radio frequency (rf) electric field in a manner that drives the ions/chemicals in a vertical up–down motion to etch
away the surface; this is portrayed in Figure 2.11(a). Unprotected metal surfaces are
removed by the process, but the resists protects any metal underneath. This results in the
cross-section shown in Figure 2.11(b). Removing the photoresist gives the final result: a
patterned metal layer that has the same features as those designed on the computer. This is
shown in Figure 2.11(c).

Figure 2.10:
After development

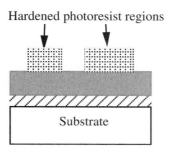

Figure 2.11:
Etching process

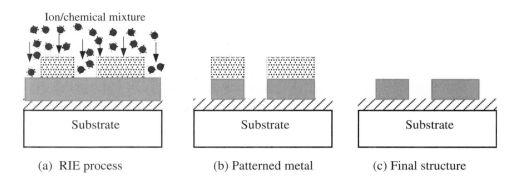

(a) RIE process   (b) Patterned metal   (c) Final structure

All material layers above the substrate are patterned using this sequence. This includes all metals and oxides (for holes). We will also need to modify some regions on the silicon substrate to make the transistor switches. To accomplish this, we use the same procedure to transfer the pattern, but change the processing steps to create the desired structures. This will be discussed later when we examine the entire process flow for an integrated circuit.

Now that we understand how the printing process works, let us take a short trip to the world of chip manufacturing to see how the process is modified for mass production. Integrated circuits are fabricated on large silicon wafers that are 200–300 mm (8–12 inches) in diameter (or larger), and about 1 mm thick. Many ICs are simultaneously created on a single wafer, as shown in Figure 2.12. The number of individual devices (or *die*) depends upon the diameter of the wafer and the shape and area of each die site. Obviously, a larger diameter wafer allows us to manufacture more ICs per wafer, but increasing the diameter makes uniform processing more difficult to achieve. When a reticle is used to pattern a layer, each die site is treated separately. The optical printing system remains stationary, and a pattern is printed at each die site by moving the wafer with highly accurate displacement motors. This is called the **step-and-repeat** process, and the piece of equipment that provides this action is called a **stepper**.

It is worthwhile mentioning that photoresist is sensitive to ultraviolet (UV) light, which is on the short-wavelength side of the optical spectrum. It does not react to long-

Figure 2.12:
Silicon wafer with individual die sites

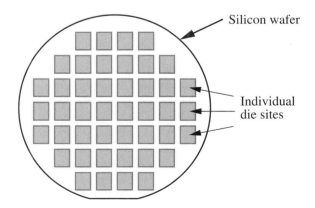

wavelength light, and it can be exposed to yellow light without causing any problems. Because of this, the printing procedure is usually confined to a "yellow-light area" in the fabrication facility. This allows us to see and maintain the operation without ruining the outcome.

Now that we have seen the photolithography process in detail, it is time to introduce some numbers into the discussion. We have talked about defining patterns in general terms, but have not specified actual sizes yet. On-chip feature sizes of modern integrated circuits are specified in units of **micrometers** denoted by μm; we often call this unit a **micron** for short. By definition,

$$1 \ \mu m = 10^{-6} \ m = 1/1,000,000 \ m = 10^{-4} \ cm$$

To get an idea of how small this really is, use a meter stick to see a 1 millimeter (mm) increment; note that $1 \ mm = 10^{-3} \ m$. Next, try to visualize dividing up the 1 mm segment into 1000 pieces; if you can, then each segment will be $1 \ \mu m$ long!

The **minimum feature size** of an IC fabrication line specifies the smallest dimension that can actually be transferred to a chip. This is determined by the resolution of the lithographic equipment and the etching processes. Smaller line geometries mean higher circuit density, so that there is a continuous drive to reduce the sizes of chip features. Modern processing facilities can manufacture chips with minimum feature sizes smaller than 1 micron; this defines the **submicron** region of silicon integrated circuits. Advanced state-of-the-art fabrication plants can now produce integrated circuits where the smallest feature is less than $0.1 \ \mu m$ wide, driving us into the **deep submicron** region.[4] The importance of these sizes will be seen when we progress to the details of layout design. Procedures and rules change as we shrink to smaller and smaller processes, and we must ensure that we follow these changes very carefully.

One thing that does not change in the lithography is that a single speck of dust on the photoresist during the exposure step can interfere with the light absorption and destroy the patterning. A single defect can render an entire chip non-functional, so this becomes a severe problem if not dealt with at the beginning. Critical chip-manufacturing steps are carried out in a **clean room** environment where the air is continuously filtered to remove particulate matter. Employees must go through a rigorous cleaning procedure before entering these areas and are required to wear special gowns that cover most of their bodies. In the most critical areas, persons will wear "bunny suits," whose only direct openings are for the eyes.

● ● ● ● ● ● ● ● ● ● ● ● ● ● ● ● ●

2.4 Planarization

Integrated circuits consist of many patterned layers of materials. After a metal layer has been patterned, a layer of insulating oxide will be deposited on top of it, followed by

---

4. "Deep submicron" usually refers to processes with minimum feature sizes less than about 0.25 microns.

Figure 2.13:
Origin of the
planarization
requirement

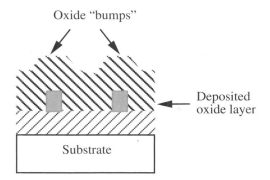

Oxide "bumps"

Deposited
oxide layer

Substrate

another layer of metal. This introduces surface "bumps" in the regions where the metal has been kept, as shown in Figure 2.13. The problem gets increasingly worse with every layer and may make it impossible to reliably form a metal pattern on the upper levels.

Planarization is a general technique developed to provide flat surfaces for every metal layer. The idea is quite simple. After an oxide is deposited over a metal, we microscopically "sand" the surface using various physical and chemical processes to give a flat surface. Once this is accomplished, we can deposit another layer of metal without worrying about the local terrain of the surface. The procedure is summarized in Figure 2.14.

Planarization is mandatory in chips that use more than about two or three metal layers. The most common technique is **Chemical-Mechanical Polishing (CMP)**, which is capable of producing very flat surfaces. All of the process examples that we will examine in this book employ planarization.

• • • • • • • • • • • • • ••••

2.5   Electrical Characteristics

Now that we have an idea of how patterns are formed, let us examine some of the important electrical characteristics that will affect the design of an integrated circuit.

Figure 2.14:
Planarization

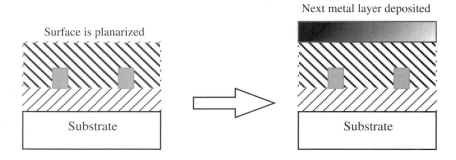

Surface is planarized

Substrate

Next metal layer deposited

Substrate

Figure 2.15:
Line geometry

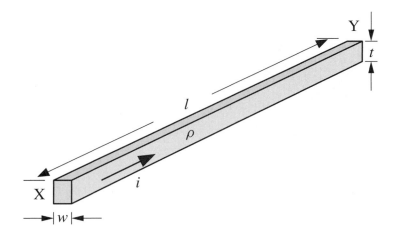

All materials have electrical resistance, *R,* that resists the flow of electrical current. If we construct a patterned line with the geometry shown in Figure 2.15, then the resistance from X to Y is given by

$$R = \frac{\rho l}{A} \ \Omega \tag{2.1}$$

where

ρ is the resistivity of the material in Ω-cm

l is the length of the line in cm

$A = wt$ is the cross-sectional area in cm^2

The resistivity ρ, depends on the material, with conductors such a aluminum (Al) and copper (Cu) having the lowest values. The thickness t of the material is determined by the processing recipe, and the chip designer does not have control over it. The main variables are the width w and the length l. These are also the dimensions seen from a top-view drawing, as portrayed earlier in Figure 2.6(b). The significance of the line resistance is that a current i flowing in the line gives a voltage

$$V_{\text{XY}} = iR \tag{2.2}$$

between the front and back. Since voltages are used to represent signals and information, the unwanted (**parasitic**) **resistance** will affect how well the physical circuit works.

Now note that conducting metal lines are patterned on top of the insulating-oxide layers. This creates a line capacitance C with respect to the silicon substrate. The geometry is detailed in Figure 2.16.

Figure 2.16:
Geometry for
capacitance
calculation

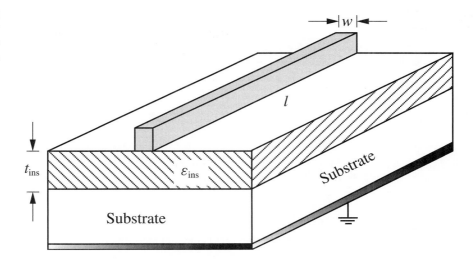

Using the simple, parallel-plate capacitor formula gives

$$C \approx \frac{\varepsilon_{ins}\omega l}{t_{ins}} \text{ Farads (F)} \tag{2.3}$$

In this formula,

t_{ins} is the thickness of the insulating oxide in centimeters (cm)

ε_{ins} is the permittivity of the oxide in Farads per centimeter (F/cm)

This ignores fringing electrical fields, so it is only a first-order estimate.

The existence of line capacitance means that we must add an electrical charge to one of the plates in order to change the voltage. Using the capacitor current–voltage (I–V) relation

$$I = C \frac{dV}{dt} \tag{2.4}$$

shows that changing the capacitor voltage by an amount ΔV requires a time interval

$$\Delta t = \frac{C}{I}\Delta V \tag{2.5}$$

This indicates a delay in the transmission of electrical signals.

Figure 2.17:
Simple *RC*
line model

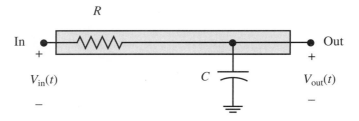

The earlier equations show that every interconnect line is associated with parasitic resistance and capacitance. A simple model of the line with these elements is shown in Figure 2.17. If we apply a step voltage to the input side of the form

$$V_{in}(t) = V_1 u(t) \tag{2.6}$$

where V_1 is a constant and $u(t)$ is the unit step function

$$u(t) = \begin{array}{l} 0 \ (t < 0) \\ 1 \ (t \geq 0) \end{array} \tag{2.7}$$

then the input voltage represents a binary transition from 0 to 1. The output voltage taken across the capacitor is given by

$$V_{out}(t) = V_{DD}[1 - e^{-t/\tau}] \tag{2.8}$$

where $V_1 = V_{DD}$, the maximum voltage level and

$$\tau = RC \tag{2.9}$$

is the time constant with units of seconds. The waveforms in Figure 2.18 show that the line introduces a delay into the signal transmission. When $t = \tau$, the output voltage has a value of

$$V_{out}(\tau) = V_{DD}\left[1 - \frac{1}{e}\right] \approx 0.63\, V_{DD} \tag{2.10}$$

Depending on the circuitry, this may not be high enough to represent a logic 1 voltage yet. Since voltages represent signals, this shows that line parasitics slow down the overall transmission of signals through a system.

It is also important to note that both R and C increase with length. Taking the product of the two gives the time constant in the form

$$\tau \approx \left(\frac{\varepsilon_{ins}\rho}{t_{ins}\, t}\right)l^2 = Bl^2 \tag{2.11}$$

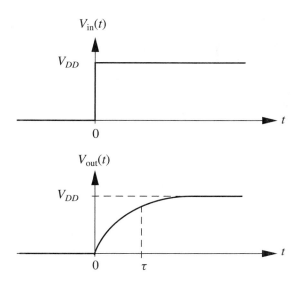

which demonstrates that the delay increases with the square of the length, and B is a constant. This implies that interconnect lines must be designed and routed very carefully, so as to not induce excessive delays in the dataflow.

Example 2.1

Consider a line that has the following dimensions: $t = 0.6$ μm, $w = 0.35$ μm, and $l = 40$ μm. The resistivity is known to be $\rho = 4.0$ $\mu\Omega$-cm. The line resistance is

$$R = \frac{(4\times10^{-6})(40\times10^{-4})}{(0.6\times10^{-4})(0.35\times10^{-4})} = 7.62 \ \ \Omega \tag{2.12}$$

To compute the capacitance, let us assume that the line runs over an oxide that is 1 μm thick and has a permittivity of $\varepsilon_{ins} = 3.54 \times 10^{-13}$ F/cm. Then

$$C = \frac{(3.54 \times 10^{-13})(0.35 \times 10^{-4})(40 \times 10^{-4})}{(1 \times 10^{-4})} \tag{2.13}$$

$$= 4.956 \times 10^{-16} \ \text{F}$$

Although these values may seem very small, they are typical of the orders of magnitude in a submicron CMOS circuit.

We note that the standard metric unit for integrated capacitance is 1 femtoFarad (fF) with 1 fF $= 10^{-15}$ F. The line capacitance is then $C = 0.496$ fF.

2.6 Silicon Characteristics

Although we have only looked at metal conducting lines, silicon is the base material for integrated circuits.

A digital integrated circuit is created by using several layers of patterned conducting materials to control the flow of signals on a small piece of silicon, called a **die**. While many layers are metal interconnect that act as microscopic wires, some layers are made with different types of crystal silicon (represented by the elemental symbol Si) which are **semiconductors**. Silicon is a very special type of material that allows us to create a type of electrically controlled switch that can be closed or opened using a control signal. In modern chip design, the switches are called **field-effect transistors (FETs).** FETs provide the basis for making logical decisions to switch (or not to switch) signals.

Silicon is called a semiconductor because it is a "partial" (or weak) conductor of electricity. It is possible to force a current to flow through silicon, but it only supports relatively small levels, compared to a conductor such as aluminum or copper. Metals can easily handle current flows of amperes, while semiconductor currents are usually measured in units of microamperes (1 μA = 10^{-6} A) or milliamperes (1 mA = 10^{-3} A). The magnitudes of the current really are not important. What makes silicon (and all other semiconductors) important is the fact that we can add small amounts of other types of atoms to the otherwise pure silicon, which alters how that region conducts electricity. The added impurity atoms are called **dopants**. This results in two distinct "types" of silicon, which in turn, forms the entire basis of microchip design as it allows us to create the FET switching devices.

If we add arsenic (As) or phosphorus (P) in very small amounts to the silicon, we find that the dopants enhance the number of free electrons that are available for conduction. The doped material thus conducts electricity better than it did in a pure silicon state. These dopants are called **donors** because each As or P atom added to the silicon increases the number of free electrons that carry electrical current; in other words, each atom "donates" an electron to the electrical conduction process. Each electron has a negative charge denoted by $-q$ where $q = 1.602 \times 10^{-19}$ Coulombs (C) is the fundamental charge unit. The resulting region of silicon is called **n-type** since it has an excess of negative electron charge carriers. In our drawings, an n-type region is indicated by simply labeling it as shown in Figure 2.19(a).

Figure 2.19:
Labeling of doped silicon regions

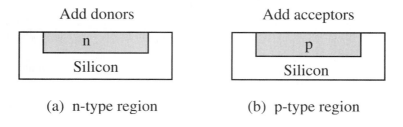

(a) n-type region   (b) p-type region

The number of donor atoms added to the sample is represented by the symbol N_d, which has units of numbers of donors per cm^3; the strict units of N_d are cm^{-3}. If we let n be the number of free electrons per cubic centimeter (also called the free electron density), then

$$n \approx N_d \qquad (2.14)$$

for an n-type material. In this approximation, every arsenic or phosphorus atom donates one free electron to the crystal. If N_d is very large, then the region is labeled as n+; this usually requires N_d to be greater than about 10^{18} cm^{-3}.

The other class of dopants used in silicon processing is elemental boron (B). When boron is added to the silicon crystal, we find that every boron atom adds a mobile positive charge called a **hole** to the conduction process. A hole is the inverse of an electron; it acts like a particle with a charge $+q = +1.602 \times 10^{-19}$ C, but it still contributes to the conduction process. This creates a **p-type** region and is labeled as shown in Figure 2.19(b). A hole can "absorb" an electron and neutralize it, so we call boron atoms **acceptors** since they can "accept" electrons. It is worth noting that this process is called **electron-hole pair annihilation**, such that

$$1 \text{ electron } (-q) + 1 \text{ hole } (+q) = 0 \text{ Coulombs} \qquad (2.15)$$

Denoting the number of boron acceptors per cubic centimeter by N_a, the number of positively charged holes, p, per cubic centimeter is given by

$$p \approx N_a \qquad (2.16)$$

This is called a p-type material since there is an excess of positively charged holes in the region. If Na is very large, then the region is labeled as being p$^+$.

A fine point in this discussion is that an n-type material still has a few holes, while a p-type sample always has some electrons. The relationship between holes, n, and electrons, p, is given by the **mass–action law**, which states

$$np = n_i^2 \qquad (2.17)$$

where n_i is called the **intrinsic density** and represents the number of electrons or holes in an undoped (pure or intrinsic) sample. In silicon at room temperature ($T = 27°$ C $= 300$ K), $n_i \approx 1.45 \times 10^{10}$ cm^{-3}. This brings us to an important point:

Electrical conduction in silicon is dominated by either electrons (n) or holes (p) depending upon the material. The resistance of a patterned silicon line is still calculated using equation (2.1). The resistivity of an n-type region is given by

$$\rho_n = \frac{1}{q\mu_n N_d} \qquad (2.18)$$

where μ_n is called the **electron mobility** with units of cm^2/V-sec. If the material is p-type, then

$$\rho_p = \frac{1}{q\mu_p N_a} \tag{2.19}$$

with μ_p being the **hole mobility**. These equations show that the line resistance varies with the doping densities. In general, the resistivity of silicon regions are much larger than that of metals, so silicon is not typically used for interconnects. The main application of doped-silicon regions is to make transistors.

One important fact from semiconductor physics is that electrons and holes behave differently in the conduction process. Electrons are said to be more mobile than holes. In terms of physical parameters, this means that μ_n is larger than μ_p in any region of a semiconductor. If we create an n-type sample using a donor-doping N_d, and compare it to a p-type sample that has an acceptor doping of $N_a = N_d$, then $\rho_p > \rho_n$. An n-type line thus has a smaller resistance than a p-type line with the same dimensions. This will also be true of the transistors that are created using the two different types of doped silicon.

The capacitance of a line formed by doping a region of silicon is more complicated, since it depends on the bordering material. When n-type and p-type regions touch, a **pn junction** is formed. This leads to a junction capacitance that affects signal delays.

We can now start to see some of the features of the chip structure evolving. Transistors are built in silicon, which forms the base substrate for the entire chip. Alternating metal and insulator layers are deposited on top of the substrate, and these are used for wiring the transistors together. The finished structure represents the physical realization of the circuit in silicon.

• • • • • • • • • • • • • • •

2.7 Overview of Layout Design

Chip layout can be characterized as the step in the design hierarchy where an electronic circuit is transferred to a silicon description. The layout designer is responsible for creating the patterns on every layer such that the resulting stacked structure defines

- The electronic switching devices (transistors)
- The wiring that connects the switching devices together

To do this, we must first examine what constitutes a transistor, and then learn the characteristics of the conducting and insulating layers. This results in a set of rules that help guide the complex task of defining every part of every transistor correctly and then providing the metal wiring.

Layout is performed entirely on a computer using a layout editor. Every layer on the chip is patterned so that the resulting layers stack into 3-dimensional structures that constitute the electronic network. A common grid is used throughout the process, and the screen portrays the entire design, but the patterning information for each layer is stored in a separate database. The data is used to create the reticle so that the patterns can be transferred to the chip during the photolithographic process.

It is the responsibility of the layout designer to insure that the patterns follow the geometrical guidelines established for the fabrication process. Violating the rules may result in a non-functional chip. The layout tools also provide the ability to translate a set of patterns into the equivalent electronic circuit for comparison. The Layout Versus Schematic (**LVS**) check insures that the patterns accurately represent the desired circuit.

The toolset is used to **extract** a circuit schematic from the layout drawings. This provides a listing of every electronic element and the wiring details; the parasitic resistance and capacitance of every line can also be determined. The extracted file is used to simulate the electronic behavior of the silicon circuit. This simple overview shows that physical design is dependent on every level of the design hierarchy. The shapes and sizes of the material layers created in the layout process determine much of the final electrical characteristics of the fabricated chip. It is therefore considered to be a critical part of the design hierarchy.

● ● ● ● ● ● ● ● ● ● ● ● ● ● ●

2.8   References

[2.1]   Campbell, S. A., *The Science and Engineering of Microelectronic Fabrication.* New York: Oxford University Press, 1996.

[2.2]   Baker, R. J., Li, H. W., and Boyce, D. E., *CMOS Circuit Design, Layout, and Simulation.* Piscataway, NJ: IEEE Press, 1998.

[2.3]   Uyemura, J. P., *A First Course in Digital Systems Design.* Pacific Grove, CA: Brooks-Cole, 1999.

[2.4]   Uyemura, J. P., *Introduction to VLSI Circuits and Systems.* New York: John Wiley & Sons, 2002.

● ● ● ● ● ● ● ● ● ● ● ● ● ● ●

2.9   Exercises

2.1   An integrated circuit line has a length of l_o and a time-constant delay of t_o. Find the delay in terms of t_o for lines of the following lengths.

    (a) $l = 2l_o$

    (b) $l = 4l_o$

    (c) $l = 8l_o$

2.2 A metal line is 0.60 μm thick and 0.25 μm wide. It has a length of 25 μm and runs over an oxide layer that is 1.2 μm thick.

  (a) The resistivity of the material is known to be $\rho = 1.8$ $\mu\Omega$-cm. Calculate the line resistance.

  (b) Find the line capacitance in farads (F) if the permittivity of the oxide is $e_{ins} = 3.54 \times 10^{-13}$ F/cm.

2.3 A silicon region is a doped n-type with $N_d = 2 \times 10^{18}$ cm^{-3}.

  (a) Calculate the electron density n.

  (b) Find the resistivity of the region if $\mu_n = 125$ cm^2/V-sec.

  (c) A patterned line is made in the n-type region. The width is 0.30 μm and the line is 9.6 μm long. The layer is 0.15 μm thick. Find the resistance.

2.4 A silicon region is a doped p-type with $N_a = 2 \times 10^{17}$ cm^{-3}.

  (a) Calculate the electron density p.

  (b) Find the resistivity of the region if $\mu_p = 80$ cm^2/V-sec.

  (c) Find the line resistance if a line is made in the p-type region with the same dimensions as in Problem 2.3(c).

CHAPTER

3

CMOS Technology—A Basis for Design

CMOS technology is the driving force behind high-density chip design.[1] It is safe to say that most of the integrated circuits in your computer are CMOS-based. The *MOS* part of the acronym stands for **m**etal-**o**xide-**s**emiconductor and originates from the basic layering scheme in the original structures developed in the 1960s. The *C* stands for **c**omplementary and originates from the circuit-design methodology. In this chapter, we will examine the processing flow used to fabricate CMOS chips and learn how to create transistors. This will provide the basis for our detailed introduction to physical design in Chapter 4.

● ● ● ● ● ● ● ● ● ● ● ● ● ● ● ●

3.1　Meet the MOSFETs

MOSFET is an acronym for **MOS f**ield-**e**ffect **t**ransistor. It is the smallest switching element that can be used on a mass-produced integrated circuit, so that much time, money, and energy has been spent designing and building smaller and faster MOSFETs. In our treatment, a MOSFET is simply a 3-dimensional structure built by sandwiching patterned regions from different material layers. Unlike the metal wiring we studied in previous sections, transistors are fabricated directly in the silicon.

There are two **polarities** or **types** of MOSFETs found in CMOS. An n-channel MOS-FET conducts electrical current using negatively charged electrons. These will be called nMOS transistors or simply nFETs in this discussion. A p-channel MOSFET (pMOS or pFET) is exactly the opposite: it conducts electrical current using positively charged particles called holes. nFETS and pFETs are said to be electrical complements of each other, and they are often used in pairs (hence the origin of the complementary in CMOS).

Let us examine an nFET first. A cross-sectional view is shown in Figure 3.1 and is similar to the figures you will see using the Microwind 2-dimensional view command.

1.　CMOS is pronounced as *see moss*.

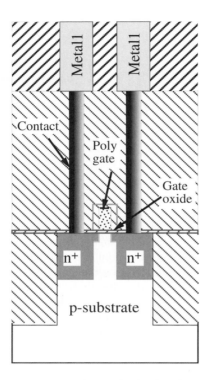

Cross-hatched areas in the drawing represent insulating oxide regions. The transistor itself is built in a silicon that is denoted by "p-substrate" in the drawing. The "p" designation means that small amounts of boron (B) have been added the silicon crystal. This increases the number of positive charges in the region, giving rise the designation "p-type" material.

The central region of the nFET is the most important. The small region marked "poly gate" is the control electrode for the device. As we mentioned earlier, "poly" is a shorthand designation for poly-crystalline silicon, which is just a conducting layer to us. The voltage on the gate determines the ability of the nFET to conduct an electrical current between the two separated "n^+" regions in the silicon. An n^+ region is one where we have a large excess of negatively charged electrons to carry an electrical current. These are created by adding arsenic (As) or phosphorus (P) to the otherwise pure silicon.

The contacts in the drawing represent metal filler that provides an electrical connection between the n^+ regions in the silicon and the Metal1 patterns at the top of the structure. They are used to gain access to the silicon-level device from the metal layers above the chip and are created by etching holes in the oxide layer, then adding the metal. Tungsten (W) is probably the most common contact material used with this approach. Copper (Cu) can also be used for contacts, but the procedure is quite different.

The last, but perhaps most important, feature of the nFET is the gate oxide between the poly gate and the silicon surface. This forms an insulating layer between the conducting poly gate and the p-substrate. In practice, the p-substrate is connected to an electrical ground, so that applying a voltage to the gate causes the structure to act like a capacitor.

Figure 3.2:
nFET top view

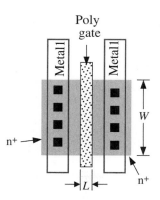

The gate oxide is very thin. Denoting the thickness by t_{ox}, a typical value in a submicron process is $t_{ox} = 30$ Å $= 3.0$ nm, where 1 Å (Angstrom) $= 10^{-10}$ m $= 10^{-8}$ cm and 1 nm (nanometer) is 10^{-9} m. Compared with a poly thickness of about 0.2–0.5 μm $= 200$–500 nm, it is really too thin to properly scale in the drawing. In early-period MOSFETs, the gate was made out of aluminum **m**etal, not poly as in modern transistors. Combining this with the gate **o**xide over the **s**emiconductor substrate gives the origin of the MOS acronym: metal on oxide on semiconductor. Even though metal gates are no longer used in practice, MOS is still used to describe this capacitor system. We will study it in more detail when we introduce the electrical characteristics of MOSFETs.

Layout of a physical design is based on the structures as seen from a top view. An nFET is drawn in Figure 3.2. No oxide layers are shown explicitly, and the entire device is assumed to be surrounded by an oxide barrier as in the cross-sectional view. Contacts are shown as black squares; note that there are several on each side of the transistor. Two new parameters have been defined. The **channel width** is denoted by W and shows the extent of the n^+ region parallel to the gate. It has practical units of microns (μm). The **channel length** L is equivalent to the width of the poly gate line, and also has practical units of microns. Both W and L are called "drawn" values because they represent the mask values that we draw on the screen. For most transistors, L will be the same as the minimum feature size. After the device is fabricated, the final (effective) values will be slightly different. The aspect ratio of the device is defined by W/L and is an important electrical design parameter that is set in the layout.

The nFET circuit symbol is shown in Figure 3.3. The symbol is used to represent the n-channel MOSFET in electrical circuit schematic diagrams. It shows the basic architectural features of the device, making it easy to translate between the symbol in circuit schematics and the physical structure. The poly gate is the control electrode that is electrically

Figure 3.3:
nFET circuit symbol

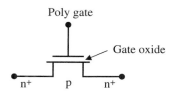

Figure 3.4:
pFET structure

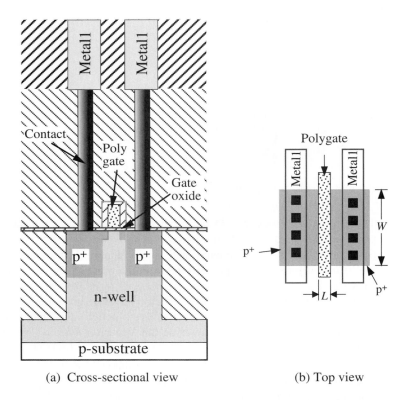

(a) Cross-sectional view　　　　(b) Top view

isolated from the main FET body by the gate oxide. The n$^+$ regions are embedded in the p-substrate.

A p-channel MOSFET is very similar to an n-channel device. In a CMOS process, nFETs and pFETs are made in the same substrate. The main difference between the two is that the n- and p-type regions are reversed. Figure 3.4(a) shows the cross-section of a simple pFET. Note that the substrate has been made into an n-type region called the n-well. Also, the left and right side of the device are p+ regions, indicating an excess of positive charges there. The top view in Figure 3.4(b) shows this change to p$^+$ regions, but the definition of the channel length L, width W, and apect ratio (W/L) are the same as for nFETs.

The pFET circuit symbol is visually the same as the nFET, except for the addition of an inverting "bubble" on the gate, as shown in Figure 3.5. The bubble is added to indicate that the pFET is the electrical opposite of the nFET; the difference in polarity gives a

Figure 3.5:
pFET symbol

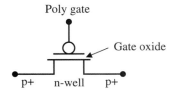

device in which the current flow and switching characteristics are "inverted" from that of an nFET.

MOSFETs are used as controlled switching devices in logic circuits. A simple view of a CMOS logic network is a collection of FETs that are wired together using metal patterns. The goal of physical design is to create the transistors and the **interconnect** wiring on the silicon substrate so that the logic is correct and the circuit exhibits fast switching. Good design requires that we have a thorough knowledge of the CMOS processing steps before we go into the details of mask design.

3.2   CMOS Fabrication

The CMOS fabrication sequence gives the step-by-step evolution of the chip from a bare silicon wafer to the finished product. Once we understand the order of the layers and how they combine to make transistors, we can progress to the details of physical design. We will use a typical submicron process to illustrate the main ideas. In practice, you should always study the specifics of the process you design in. When following the process described next, we will track each patterning step since that implies the need for a separate mask.

The starting point in our process is a p-type silicon wafer, as in Figure 3.6. The wafer is in fact a single crystal structure in which the atoms arrange themselves in a nice periodic array; crystalline silicon is required to create the transistors. The wafer is made p-type by adding small amounts of boron to a molten silicon mixture before the material solidifies. Both nFETs and pFETs need to be built in the substrate. It is useful to refer back to the cross-sectional views in Figures 3.1 and 3.4 as we progress so that you can see the structures take shape. To allow us to keep some detail in the drawings, the layer heights (thicknesses) will not be to scale.

Mask 1: n-Well

The transistor process starts by identifying the location of pFETs and adding arsenic or phosphorus to create the n-well regions. The process of adding impurity atoms, such as arsenic, to the silicon is called **doping**, and the impurities themselves are called **dopants**. To force the dopants into the silicon wafer, the atoms are ionized and then accelerated using ion guns. This process is called **ion implantation** and is the standard way to dope silicon regions. The wafer is then heated in a furnace to heal the damage created by the impact of the ions on the silicon crystal. This step is called **annealing** and is required to help the dopants set correctly into the crystal structure.

Figure 3.6:
Starting wafer

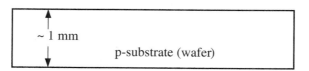

Figure 3.7:
Formation of the
n-well region for
pFETs

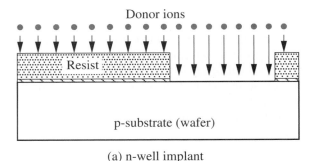

(a) n-well implant

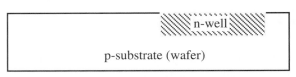

(b) Final structure

The formation of the well "layer" is shown in Figure 3.7. A mask is used to define the n-well regions. After exposure and development, a hardened layer of photoresist covers and protects all sections of the silicon surface that are not n-well. An n-type ion implant (arsenic or phosphorus) can penetrate unprotected areas and form the n-regions as portrayed in Figure 3.7(a). The resulting structure in Figure 3.7(b) shows the n-well after the annealing step is completed.

Mask 2: Active Areas and FOX

High-density integrated circuits need to pack transistors very close together on the silicon surface. To insure that adjacent devices do not have any unwanted electrical conduction paths among them, individual transistors are surrounded by glass (oxide) regions that are created in the silicon. This is a general form of device **isolation**.

Regions of a chip are divided into two categories, depending upon their usage. Transistor sections are called **active** areas, since they contain active electronic devices. Any space that is not active is defined to be a **field** region. Symbolically, we may write that

Active + Field = Chip surface

Field oxides (**FOX**) are formed to provide device isolation.

Isolation regions are defined by the **active** area mask. This mask is used to define FET regions on the surface by patterning a dual layer of silicon nitride (Si_3N_4) and oxide, as shown in Figure 3.8(a). Nitride is used because it is a very dense dielectric that protects the silicon surface. The nitride/oxide layer itself is used as a mask for the etching process depicted in Figure 3.8(b). RIE is used to form trench-like regions in the silicon to physically separate neighboring transistors. A thin layer of oxide is then grown over the surface, followed by a p^+ implant into the trench. Depositing oxide on top then fills the field

Figure 3.8:
Active area
definition and
FOX formation

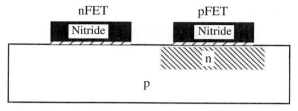

(a) Active area definition

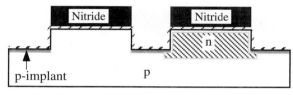

(b) Substrate etch

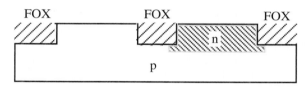

(c) Field-oxide deposition

regions, yielding the view in Figure 3.8(c). The key to this type of device isolation is to insert a glass barrier between neighboring devices.

At this point, separate ion implants are made to nFET regions and pFET regions to adjust the "turn-on" parameter called the **threshold voltage**. We will examine the effect of this in later chapters.

Mask 3: Polysilicon Gate

The next sequence of steps is directed towards forming the actual transistors. This starts with the growth of the gate oxide on the surface. Passing oxygen over a heated silicon wafer forms silicon dioxide according to the reaction

$$Si + O_2 \rightarrow SiO_2$$

The oxide is overgrown to insure that there are no "pinholes" in it, and then etched back to a working thickness of t_{ox}, as in Figure 3.9(a). No mask is required for this step, since the oxide is grown over the entire surface of the wafer. Some processes allow the user to selectively block the etchback, which results in FETs with thicker gate oxides. These yield devices with higher threshold voltages that can be used with higher power-supply levels.

Figure 3.9:
Gate oxide growth
and poly gate
formation

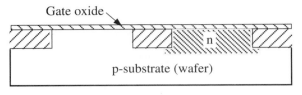

(a) Gate oxide growth

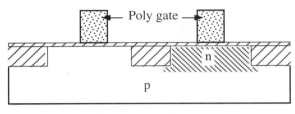

(b) Poly gate formation

MOSFETs are formed using what is called a **self-aligned** technique where the gate and n^+ (or p^+) regions automatically are aligned to each other. The gate is formed by depositing doped silicon on top of the gate oxide and then patterning the layer to define the gates; Figure 3.9(b) shows the resulting cross-section. Silicon has a natural tendency to form a crystal, but silicon dioxide (the gate-oxide layer) is amorphous, so it does not provide a good base. Because of this, the silicon forms small regions of crystals, and the overall material is called **polycrystalline** silicon, or **poly** for short. For modeling purposes, we usually treat it as being similar to regular silicon regions.

Masks 4 and 5: n^+ and p^+ Doping

The next steps are ion implants that are used to define the p^+ and n^+ regions of pFETs and nFETs, respectively. The "+" designation of p- and n-type regions is used to imply that the doping density is very large. In device terminology, the regions form what are known as the **drain** and **source** terminals of a FET, so that they are also called the drain/source masks. In this sequence, we will create what is known as a **lightly doped drain (LDD)** MOSFET. These structures are used to reduce some small device effects due to very energetic particles called **hot electrons** and **hot holes**.[2]

Let us start with the p-type boron implant use to form pFETs. The first step is a "light" boron implant that creates p^- (lightly doped) regions in the pFET, as shown in Figure 3.10. The resist mask blocks the boron from entering nFET sites and concentrates the dopants over pFET regions. The process is called "self-aligned" because the poly gate itself acts as a mask to the incident ions, blocking them from entering the silicon below. This gives p^- silicon doping on the left and right side of the gate, but the silicon directly under the gate is still n-type. An nFET structure is created in the same manner. A mask is used to create a

2.   In this context, "hot" refers to a particle's energy as related to a thermodynamic temperature.

Figure 3.10:
First p-type
pFET implant

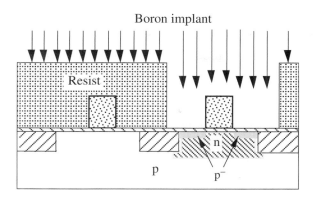

resist block over pFET regions while providing access to nFET sites. Ion implanting donors (arsenic or phosphorus) creates lightly doped n^- drain and source regions, as shown in Figure 3.11.

LDD MOSFETs combine the lightly doped drain and source regions described above with heavier p^+ and n^+ sections that make the material more conductive. Fortunately, the next steps do not require a separate mask, making the process very attractive from an economic viewpoint. The key to finishing the LDD structure is to first create **spacer** oxides on the poly gates. The sequence is shown in Figure 3.12. Oxide is deposited over the surface of the wafer as illustrated in Figure 3.12(a). The oxide is then etched vertically downward. From the top, the oxide on the sidewalls of the poly gates is thicker than other regions, so it doesn't etch all the way down. The remaining sidewall of oxide shown in Figure 3.12(b) are the spacers.

The heavy drain/source doping is completed using the spacers and poly gates as masks. Figure 3.13 portrays both the pFET [in (a)] and nFET [in (b)] implants and resulting structures. This step completes the formation of the transistors.

Figure 3.11:
Initial n-type
nFET implant

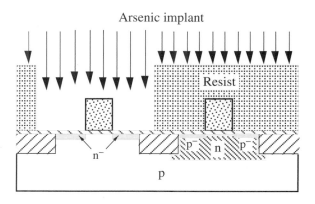

Figure 3.12:
Spacers

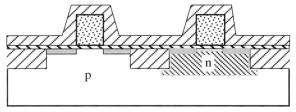

(a) Oxide deposition

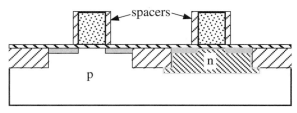

(b) Etchback and spacer formation

Figure 3.13:
Deep p+ and
n+ FET implants

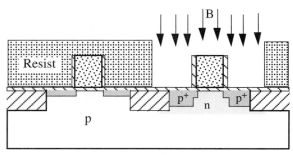

(a) Deep p+ implant

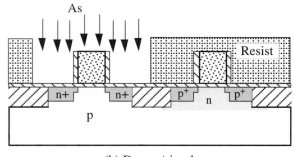

(b) Deep n+ implant

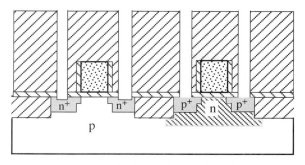

(a) Oxide deposition and etching

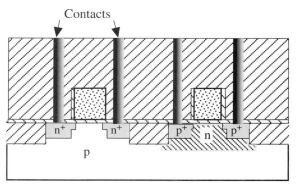

(b) Contact plug fill

Mask 6: Oxide Contacts

Once the FETs are fabricated, we must provide a means to wire them into the circuit. This is accomplished by alternate oxide–metal layerings, where the metal layers are patterned to provide the signal flow paths. The first step is shown in Figure 3.14(a) where an oxide layer is deposited uniformly on the FETs, and contact cuts (holes) are patterned in the insulating material. These are filled with conducting plugs, yielding the cross-section in Figure 3.14(b).

Mask 7: Metal1

The Metal1 layer is deposited on top of the oxide and then patterned, giving us the structure illustrated by Figure 3.15(a). This serves as the base for higher metal layers that are added by continuing the process. The next step is to deposit an oxide layer on Metal1 and then planarize the top as shown in Figure 3.15(b).

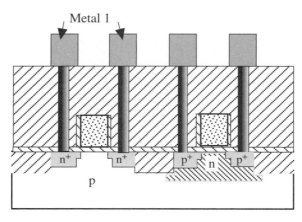

(a) Metal1 deposition and patterning

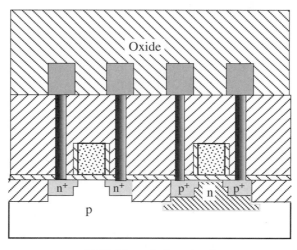

(b) Oxide and planarization

Masks 8+: Vias and Higher-Level Metal Layers

The oxide on top of Metal1 is then patterned with contact holes and filled with plug material to allow electrical contact to higher metal layers. Connections between metals are usually called **vias**. The next metal (Metal2) is then deposited, patterned, and covered with oxide. The oxide over Metal2 is patterned and then is itself covered with oxide. Continuing this process gives a cross-section like that shown in Figure 3.16.

Figure 3.16:
Higher-level metal
layers and vias

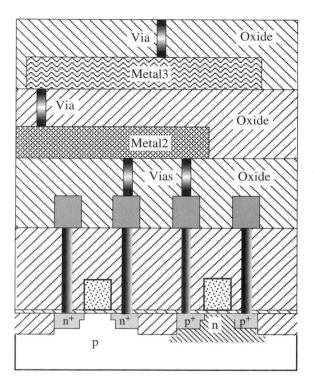

Final Masks: Metal_Top and Passivation

Let us call the top conducting layer Metal_Top. It is sometimes made extra thick, particularly if integrated on-chip inductors are included in the process capabilities. Metal_Top is covered with a layer of of silicon nitride that is used as the **passivation** coating. Nitride is a very dense dielectric that protects the chip from contaminants in the atmosphere. This is important for reliability considerations, as foreign substances may lead to chip failure if they diffuse into the structure. The passivation mask is used to open holes in the nitride that give electrical access to the top metal layer. In other words, passivation cuts define the entrance and exit points to the outside world. Figure 3.17 shows the Metal_Top and passivation coatings. This completes the chip fabrication sequence.

After the wafer is through the process cycle, it is put into a testing apparatus that probes each die. Testing is a very important, but expensive, part of the manufacturing sequence. Every die is powered up and subjected to various input combinations whose results can be observed at specified outputs. Not every die is functional or meets the specifications. Those that fail are marked and eventually disposed of. When the wafer test is

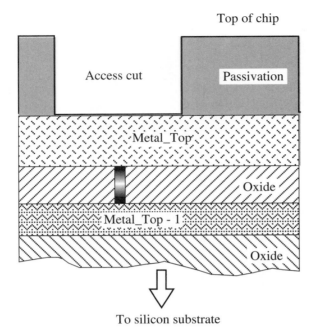

Figure 3.17:
Top metal and
passivation layer

Top of chip

Access cut

Passivation

Metal_Top

Oxide

Metal_Top - 1

Oxide

To silicon substrate

finished, a diamond-tipped blade is used to score the wafer between the die sites. A small amount of pressure cracks the crystal along the score lines, and individual die can then be mounted in packages. Additional testing may be required for the packaged devices.

3.3 Submicron CMOS Processes

CMOS processing has advanced at a remarkable rate in the past few years. Minimum feature sizes smaller than 0.1 μm are common, and new advances appear poised to continue the trend of the "incredible shrinking MOSFET." Many factors contribute to this trend.

3.3.1 Moore's Law and Integration Levels

Early integrated circuits contained only a few transistors and other components. These were used to create basic logic functions, such as the NOT, NAND, and NOR; and early ICs were placed in metal or plastic packages with only a few input/output (I/O) pins. Modern chips have several hundred million transistors, and the number of FETs will exceed one billion in the near future. Advanced IC packages have several hundred pins, yet engineers still complain that they are "pin-limited" in their designs. Of course, this massive change didn't take place overnight.

Figure 3.18:
Example of
Moore's Law

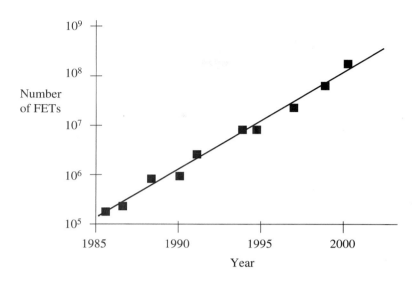

Gordon Moore, one of the founders of Intel Corporation, made a prediction that has since become known as "Moore's Law." It states that

The number of transistors on a die will double every 18 months.

At the time of his statement, the integrated circuits industry was experiencing moderate growth, but even then it seemed very optimistic. Few believed that it would extend into the 21st century. But it has! Technologists often marvel at the fact that Moore's Law, which implies shrinking dimensions with a constant rate of process improvement, still seems to be in effect today. Figure 3.18 shows an example of Moore's Law for some representative chips selected from the literature. The continuous increase in the device count is easily seen.

The concept of **integration levels** was introduced to describe the increase in transistors and functionality over the years. Traditionally, chips have been divided into four categories, although the borders are not well defined. These are given by the acronyms below; note, however, that the number of transistors quoted in each level is merely a rough estimate.

- **SSI** (Small-Scale Integration) with less than about 1,000 transistors. SSI includes chips with individual logic gates.
- **MSI** (Medium-Scale Integration) to around 100,000 transistors. An example of an MSI chip is a 4-bit arithmetic logic unit (ALU) or a basic calculator.
- **LSI** (Large-Scale Integration) to around 1,000,000 transistors. 8- and 16-bit microprocessors and basic digital signal processors (DSPs) are in this category.
- **VLSI** (Very Large-Scale Integration) to around 100,000,000 transistors. This includes the current generation of microprocessors that have about 40–50 million transistors.

Current day chip design is at the VLSI level. For circuits with a billion or more FETs or logic gates, the next level is

- **ULSI** (Ultra Large-Scale Integration) with about one billion transistors,

which some have coined

- **GSI** (Giga-Scale Integration) as an alternate.

CMOS VLSI has been the driving force in computer hardware, digital signal processing, telecommunications, and a host of other fields.

3.3.2 Yield

As mentioned in the previous section, not every die on a wafer is functional. Wafers are usually processed in large groups called lots. Every wafer in the lot is subjected to the same processing steps, so it is meaningful to trace a lot through the fabrication sequence to see how good the processing is.

The **yield** Y of a process is defined as

$$Y = \frac{\text{Number of good die}}{\text{Total number}} \times 100 \tag{3.1}$$

and gives the total percentage of good die. Obviously, a high yield is desirable as each bad die subtracts from the revenue. When a new chip is put into production, the yield may be very low because the design needs to be improved. However, as the design is modified and the recipes are refined, the yield increases to a point where it is profitable to manufacture and sell it.

Many factors contribute to the yield. One of particular interest to us is the fact that the yield decreases with increasing die area. The simplest estimate to this dependence is

$$Y = e^{-DA} \times 100\% \tag{3.2}$$

where A is the die area in cm^2 and D is called the **defect density** in units of defects per cm^2. The defect density describes point defects on the wafer surface and is significant because a single defect can ruin the circuitry on a die. Typical values of D are less than about one defect per cm^2. Figure 3.19 shows a plot of yield Y as a function of die area A. This shows that smaller die areas result in higher yields. While better yield analysis is possible, this rule generally remains valid.

Increased yield is one factor that drives the industry to smaller and smaller feature sizes. Decreasing the size of a chip results in a higher profit margin and also allows more complex systems to be integrated.

Figure 3.19:
Yield as a function
of die area

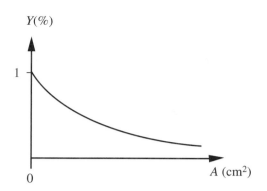

3.4 Process Technologies in Microwind

Microwind has been structured in a manner that allows different processes to be accessed and studied. This is accomplished using the selection of a **foundry** for a design.

In semiconductor terms, a foundry is a central chip-fabrication plant that accepts designs from the outside world and uses the design files to manufacture the chips. This allows design companies to create proprietary chips without having to invest several billion dollars into a manufacturing facility. In educational circles, access to various foundries is most easily accomplished using MOSIS.[3] This organization was originally established by the U.S. government for the purpose of providing government contractors and universities with a means to fabricate test chips.

Foundries characterize different process lines by the minimum allowed feature size. Other specifications are also involved, such as the addition of another polysilicon layer (Poly2) or the ability to do high-speed analog circuits. However, the minimum resolution is usually the key parameter for a process. It generally implies the smallest dimension for the drawn channel length L in a MOSFET and has a great effect on the characteristics of the circuit. All pattern dimensions are scaled with the feature size, so that the interconnect wiring provided by metal layers depends upon the specification.

To see the different foundry choices available in Microwind, launch the program and execute the Menu command sequence

File $\Rightarrow$

    Select Foundry

as shown in Figure 3.20. The keyboard sequence **[Ctrl][F]** may also be used. Either brings up the Foundry window shown in Figure 3.21. This lists several different options that may

---

3. MOSIS is an acronym for **MOS I**mplementation **S**ervice.

Figure 3.20:
Foundry access

Figure 3.21:
Foundry selection
menu

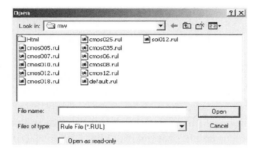

be used as a design basis.[4] The file names are written so as to imply the minimum feature size of the process. For example,

cmos012.rul

specifies a 0.12 μm minimum feature size, and it is the default technology that is loaded when Microwind is lauched. The file

cmos025.rul

is for a 0.25 μm process. The default can be set to any technology.

It is important to remember that the number of metal layers varies among the different processes. The 1.2 μm process has only Metal1 and Metal2 interconnect, while the 0.25 μm design can access 6 metal layers. If you select a specific technology file, you will be returned to the original screen and the technology and number of metal layers will be listed below the work area. Shrinking the feature size corresponds historically to improved technologies, and there are many other differences that will be found as one progresses from the 1.2 μm to the deep submicron levels.

---

4. The actual list of processes may be different, depending upon the version of Microwind that you have.

In a realistic design environment, scaling a process down to smaller geometries is much more involved than just reducing the feature size. Electrical parameters change, as do the characteristics of the FETs. And some effects that weren't noticeable in "big" geometries become major problem areas as the dimensions are reduced.

A chip designer must pay attention to these differences as the technology advances. Our philosophy will be to provide a general background in chip layout and simulation that can be extended to various technologies. We will always specify the technology that we are working in.

● ● ● ● ● ● ● ● ● ● ● ● ● ● ● ● ●

3.5   Masks and Layout

The process flow described in the previous section illustrates the relationship between the mask patterns and the layer sequence. Physical design revolves around designing a set of masks that result in 3-dimensional substructures, such as FETs, and the associated wiring on the chip.

The masking order in the basic process was as follows:

- n-well
- Active
- Poly
- p-implant
- n-implant
- Contact
- Metal1
- Via
- Metal2
- Via

  ⋮

- Metal_Top
- Passivation

where oxide layers are deposited and patterned by the Contact and Via masks. Chip layout requires that every mask be patterned, so that the 3-dimensional layered circuit creates all of the transistors and wires them together correctly. Although this may sound somewhat complicated, it is simplified by the observation that masks can be divided into two categories: FETs and interconnect. This allows us to study the two separately.

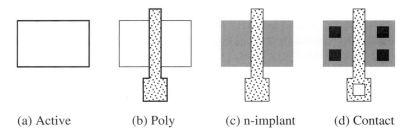

(a) Active   (b) Poly   (c) n-implant   (d) Contact

3.5.1 FET Structures

Three-dimensional MOSFET structures are formed using distinct masking groups. Reviewing the processing sequence shows that the following patterned layers are needed for an nFET:

- Active
- Poly
- n-implant

This sequence is illustrated by the top-view drawings in Figure 3.22. This shows how individual masks are aligned to create a 3-dimensional stack with the nFET layering structure. The sequence for a pFET is similar, and consists of the masking steps:

- n-well
- Active
- Poly
- p-implant

The top-view masking sequence for a pFET is shown in Figure 3.23. Note that we have included contacts to the n^+/p^+ regions and the poly gate to allow for connections to the rest of the circuit. In both devices, the drawn channel width W is set by the vertical dimension

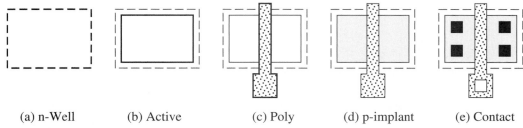

(a) n-Well   (b) Active   (c) Poly   (d) p-implant   (e) Contact

Figure 3.23: pFET masking sequence

of the Active mask, while the drawn channel length L is the given by the horizontal extent of the Poly mask.

We can generalize the approach symbolically by defining a masking layer in the abstract form

$$\text{Mask\_Pattern} = \text{Layer} + \text{Not\_Layer}$$

In this expression, **Layer** represents the geometrical features that are created by the mask. For example, **Active** would denote the collection of all active areas on the chip, while **Not_Layer** are the field regions. Similarly, **Poly** represents the presence of a poly line. Using set notation,

$$(\text{Layer1}) \cup (\text{Layer2})$$

indicates the **union** of features where either **Layer1** or **Layer2** patterns exist. This is logically equivalent to the OR operation. The **intersection**

$$(\text{Layer1}) \cap (\text{Layer2})$$

is where patterns from both layers exist, and is equivalent to the AND logic operation.

Using this symbology, we can define regions by masking layers in a compact manner. We start with doped semiconductor sections. An n^+ region will be denoted as ndiff and is created by

$$\text{ndiff} = (\text{Active}) \cap (\text{N-implant}) \cap (\text{Not\_Poly})$$

A p+ region is labeled as pdiff, and is created by

$$\text{pdiff} = (\text{Active}) \cap (\text{P-implant}) \cap (\text{Not\_Poly})$$

Note that ndiff and pdiff are derived from other masking layers.[5] Derived masks do not need to be designed separately since they can be created from existing data. By these definitions, both ndiff and pdiff are created in active regions of the p-substrate. To build an nFET, we combine the masks as

$$\text{nFET} = (\text{ndiff}) \cap (\text{Poly}) \cap (\text{Contact})$$

where we have included the contact opening for completeness. Similarly, a pFET is

$$\text{pFET} = (\text{N\_well}) \cap (\text{pdiff}) \cap (\text{Poly}) \cap (\text{Contact})$$

---

5. The terms "ndiff" and "pdiff" are short for n-diffusion and p-diffusion, respectively. This terminology originates from early silicon fabrication schemes where the n^+ and p^+ regions were doped by diffusing impurity atoms into the silicon substrate using heat. Ion implantation has replaced impurity diffusion, but the terminology still remains (like MOS).

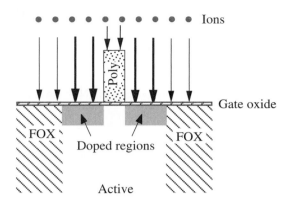

This places all pFETs in n-well regions as required in the process flow. These types of expressions help us visualize how features of each layer stack to form the 3-dimensional devices on the finished chip.

A very useful layout rule to remember arises from the self-aligned gate process. The implant step in Figure 3.24 details the formation of a doped region. Ions are incident, or impinge over the entire surface of the wafer, but can be blocked from entering silicon by either the poly gate layer or field oxide (FOX). The only setions that are doped are those where a layer of thin gate oxide is the only barrier that the ions must penetrate. This observation is used as the basis for the general statement that

Poly drawn over diffusion creates a FET.

In terms of the terminology introduced above, we can be a bit more specific and say

Poly over ndiff gives an nFET.

Also,

Poly over pdiff in n-well gives a pFET.

These rules help us remember how to design FETs in layout drawings. It also establishes some rules of layering in that these are always true, so that you must be careful when drawing poly patterns around ndiff and pdiff if you do not want to inadvertently form a transistor.

3.5.2 Interconnects

The remaining masking layers in the basic CMOS process are dedicated to wiring the circuits together. The conducting layers are metals with insulating oxides inbetween. The important masks are:

- Contact
- Metal1
- Via
- Metal2
- Via
  ⋮
- Metal_Top

Metal masks provide the wire routing, while contacts and vias allow us to move between conducting layers to make the routing problem easier.

Routing requirements are based on the design of the circuits that are implemented on the chip. Some are simple, while others can be quite complicated. As we will see later, interconnect wiring differs greatly as we move from small, simple circuits to large, complex systems.

● ● ● ● ● ● ● ● ● ● ● ● ● ● ● ⋯

3.6   The Microwind MOS Generator

Now that we are familiar with the CMOS fabrication sequence, let us use Microwind to create our first transistor layouts.

Launch Microwind and examine the screen. The Palette window appears. The top portion of the Palette menu is shown in Figure 3.25. Notice the button in the lower-left side of the screen shot that looks like a FET circuit symbol. This provides access to the MOS generator routine in Microwind. Simply place the cursor on the button as shown and left-click the mouse to bring up the dialog screen. Alternately, you can use the menu commands

Edit ⇒
   Generate⇒
      nMOS

which brings up the same screen.

Figure 3.25:
MOS Layout
Generator button

Figure 3.26:
MOS Layout
Generator
dialog screen

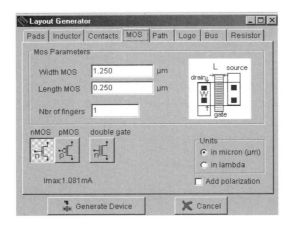

The MOS Layout Generator has several options, all of which are controlled by the screen shown in Figure 3.26. The channel length L is preset by the selected process; in the example, it is 0.25 μm. The FET channel width W also is preset to 1.250 μm. Your values may be different depending upon the program version you are using. Normally, the channel length will be left at the default value, but the designer adjusts the value of W according to the circuit specifications. In most cases, it is the value of W/L that is the important parameter. We will leave both at their default values for now. If you are following the steps on a computer as you read (highly recommended!), just use any values that are on your screen.

To create an n-channel transistor, push the nMOS button if it is not already selected; the screen dump shows the appearance after it has been pushed. Then push the Generate Device button. This action will place you back into the drawing screen. Specify the location where you want the nFET by positioning the cursor, then left-click the mouse. An nFET with the specified width and length will appear in the work area. Figure 3.27 shows an edited monochrome screen dump of the device, but yours will be color-coded according to the scheme

- Poly ⇔ red
- n^+ ⇔ green

Figure 3.27:
nFET using the MOS
Layout Generator
dialog screen

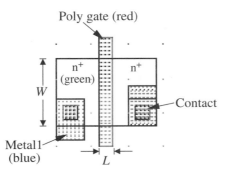

Figure 3.28:
Generated pFET

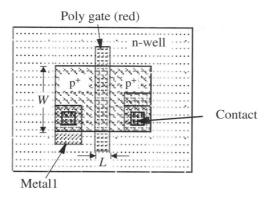

- Metal1 ⇔ blue
- Contact ⇔ box with white X

The W and L values indicated in the figure are scaled according to an internal ruler; we will examine the scaling in the next chapter. For now, the important aspects of the FET are the color-coding for the layers.

A pFET can be generated by the same technique, except that you push the pMOS button in the dialog screen of Figure 3.26. Positioning the cursor on the work screen in a blank area and left-clicking creates a pFET, as in the screen dump of Figure 3.28. The outline of the FET is similar, except that it uses a brownish-orange (depending upon your monitor settings) to indicate the p^+ regions. Also note that the pFET is embedded within an n-well that is color-coded as a green rectangle with a green stipple pattern fill.

The last feature of the MOS generator we will study here is the Add polarization box in Figure 3.26. In the terminology of Microwind, "polarization" means to add a positive or negative bias voltage. If you select this option by clicking the box, both the generated nFETs and pFETs are modified to those shown in Figure 3.29. The nFET has an added section of Metal1 and a contact to the substrate; this is used to insure that the p-substrate

Figure 3.29:
Polarization option
in the MOS Layout
Generator

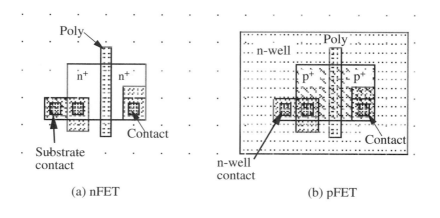

(a) nFET   (b) pFET

has a well-defined voltage and is not a floating value. In practice, this is most commonly set at 0 V (ground). Similarly, the generated pFET device now has a contact to the n-well, which is added for the same purpose; in our designs, the n-well will be electrically connected to the positive external power supply.

These examples illustrate the main connection between the CMOS fabrication process and mask design and layout. The MOS Layout Generator can be used to create the FETs needed for a circuit, and the transistors are wired together using metal layers with contacts and vias. Of course, it is not quite that simple, because we must follow a strict set of rules and guidelines to create a functional circuit. Learning the techniques is not difficult, however, and is often considered to be one of the most enjoyable parts of chip design. Even fun!

● ● ● ● ● ● ● ● ● ● ● ● ● ● ● ●
3.7 Chapter Summary and Roadmap

We have studied the CMOS fabrication sequence to understand how the layers are stacked to form transistors and interconnect wiring levels. Each layer is designed on a separate mask, and the resulting patterns stack in the chip to form 3-dimensional physical structures. The masks are thus the key to designing a chip.

The next chapter introduces you to the concepts involved in using a layout editor as a CAD tool. Microwind is used as a vehicle to teach the drawing and editing of masking patterns. This provides a connection to the CMOS fabrication sequence discussed in this chapter. Chapter 5 introduces the "rules of layout design" that govern the way we design masks for the CMOS process. The material in Chapter 6 and beyond uses the fundamentals to take you deep into the design of CMOS chips.

● ● ● ● ● ● ● ● ● ● ● ● ● ● ●
3.8 References

[3.1] Baker, R. J. H., Li, W., and Boyce, D. E., *CMOS Circuit Design, Layout, and Simulation*. Piscataway, NJ: IEEE Press, 1998.

[3.2] Clein, D., *CMOS IC Layout*. Boston: Newnes, 2002.

[3.3] Smith, M. J. S., *Application-Specific Integrated Circuits*. Reading, MA: Addison-Wesley, 1997.

[3.4] Uyemura, J. P., *A First Course in Digital Systems Design*. Pacific Grove, CA: Brooks Cole, 2000.

[3.5] Uyemura, J. P., *Introduction to VLSI Circuits and Systems*. New York: John Wiley & Sons, 2002.

● ● ● ● ● ● ● ● ● ● ● ● ● ● ●

3.9   Exercises

3.1   Use the MOS Generator in Microwind to create an nFET with $W = 4L$. Use the default value for L to determine the required W.

3.2   Suppose that we want to describe an electrical connection between Metal1 and diff using abstract mask layers, as in Section 3.5.1. Write the appropriate expression.

3.3   Use the MOS generator in Microwind to create a pFET with a $W = 4L$. Use the default value for L to determine the required W.

Using a Layout Editor—Fundamental Concepts

Physical design is based on the characteristics of the process flow. A layout editor allows the designer to create FETs and the metal interconnect wiring that connects the transistors together. Every layer must be patterned so that the 3-dimensional structure functions as intended. This chapter will introduce you to the fundamental concepts of drawing and editing layers, with hands-on practice using Microwind.

4.1 Lambda-Based Layout

CMOS processes tend to be identified by the smallest feature size that can be fabricated reliably on a chip. For example, a 0.18 μm process usually is interpreted as one that can be resolved down to 0.18 microns, but should not be used to create anything smaller. Layout design is the art of drawing the mask sets that create the chip layer patterns, so that it is intrinsically related to the CMOS processing characteristics.

The resolving limits of the processing enter layout design in the concepts of the **minimum width** and **minimum spacing** requirements. The simplest geometrical object in layout is a rectangle, as shown in Figure 4.1(a). Elongating the rectangle so that one side has the smallest dimension w allowed by the processing gives a minimum width **line**; this is shown in Figure 4.1(b). Lines are used to pattern interconnect wiring that electrically links points in a circuit. When two lines of the same layer are placed next to each other, as in Figure 4.1(c), the edge-to-edge spacing s becomes important since it is also a physical feature. The smallest allowed values of w and s for a specific CMOS process are called the minimum width and minimum spacing, respectively. The numerical values should be viewed as hard limits that cannot be violated, as they represent the physical limits of the fabrication process.

Figure 4.1:
Rectangles and
lines in a layout
drawing

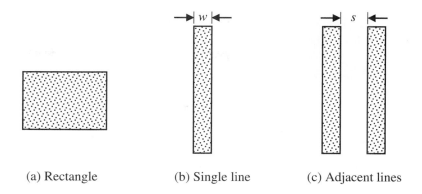

(a) Rectangle   (b) Single line   (c) Adjacent lines

Minimum width and spacing values are specified for every layer on the chip. They are part of a larger group of geometrical specifications that collectively are known as layout **design rules (DRs)**. The values originate from the limits imposed by the lithographic imaging and etching, but also incorporate electrical and mechanical effects as well. Design rules must be followed at all times during the mask design (layout) procedure. A routine called the **Design Rule Check (DRC)** is provided in the layout editor to help find DR violations that may have been missed. This allows us to fix the problem before the design is sent to the fabrication process.

There are two ways to specify design rules. The most obvious one is to assign numerical values to every important width, spacing, and specialized situation that may arise in constructing a CMOS layout. These are called **process-specific** rules. For example, we may stipulate that the minimum allowed width for a polysilicon layer is $w = 0.35$ μm and the minimum allowed spacing to the edge of adjacent poly feature, is $s = 0.525$ μm. A Metal1 line in the same process might have minimum values of $w = 0.475$ μm and $s = 0.475$ μm based on different considerations. Design rule sets of this type are very specific and allow for the maximum packing density. Most commercial chip designs are based on process-specific rules.

The alternate approach is to create a set of **scalable** design rules that allow us to construct layouts that can (in principle) be moved from one process to another by a simple procedure. To achieve this, we introduce a **length metric** λ (**lambda**), in units of microns (μm) and specify all minimum sizes in units of λ. Each process is characterized by a numerical value for λ that determines the critical minimum dimensions. For example, poly layout may be governed by the rules $w = 2\lambda$ and $s = 2\lambda$. In a quarter-micron process, $\lambda = 0.125$ μm, so that the minimum values are $s = 0.25$ μm and $w = 0.25$ μm.

Lambda (λ) design rules are very convenient if you want to compare layouts and performance of a circuit in different CMOS processes. The design can be completed without regard to a specific technology and then scaled as needed with the numerical value of λ. The price paid for this convenience is that the approach does not allow one to achieve the maximum packing density on the silicon. This situation arises because of several factors. One, is that the dimensions are stated as integer multiples of λ, such as 2λ and 3λ, which automatically eliminates physical dimensions of, say, 2.45λ that would maximize area

Figure 4.2:
On-screen
lambda ruler

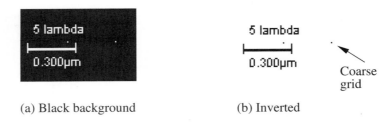

(a) Black background   (b) Inverted

usage. Another factor is that CMOS circuits have physical effects that do not scale linearly with size, and these affect the value of λ itself. For example, in a 0.18 μm technology, λ = 0.10 μm; for a poly layer designed with w = 2 λ, the smallest drawn width would be w = 0.20 μm.

Even though there are drawbacks to using scalable lambda-design rules, their portability makes them attractive for many situations such as prototyping, proof-of-concept circuits, research projects, and archiving circuits for re-use. Most academic institutions use scalable design rules since they provide a sound basis for learning layout and the resulting circuits can be fabricated in silicon without modifications. Microwind is configured as a Lambda-based layout editor which allows us the freedom to explore our designs in many different technologies. When you change the selection of foundry in the **File** submenu, your layouts are scaled automatically to the appropriate value of λ.

Example 4.1

Microwind displays a lambda metric measurement of the CMOS process in the work area. To see this, launch Microwind and examine the upper-left hand corner of the screen. The Lambda-ruler appears as in Figure 4.2(a); the inverted (white background) view is also shown in Figure 4.2(b). In this example, the indicator shows the length of a 5 λ segment for the screen. The numerical value is shown as 5 λ = 0.300 µm, so that λ = 0.06 μm for this process. The values that you see on your screen may be different, but the view will be the same. The ruler is a handy reference when you draw.

Note that there are grid points in the work area. In the launch screen, you will see the coarse grid points as being bright dots in the black background. The lambda ruler is the length of the distance between coarse grid points, i.e., 5 λ. Depending upon the contrast settings of your monitor, you may also see dimmer dots in between the coarse grid points. These can be expanded by using the **Zoom-in** command from either the menu or the keyboard sequence **[Ctrl][Z]**. Once activated, point to the region that you want to magnify and click. Alternately, you can construct a box around a selected area using the cursor and mouse button; the action is similar to drawing a box described in the next section. The lambda ruler tracks changes in the view, and the grid becomes clearer as you zoom in. If you zoom in too far, you can zoom out or use the **All** button to see the entire layout.

The scaled grid is shown in the screen dump of Figure 4.3(a) with the inverted shot in Figure 4.3(b); the white background is an option in the File Menu. The coarse grid points are brighter than the regular grid points so that you can distinguish between them. Figure 4.4 provides a drawing that better illustrates the structure and meaning of the grid. The

Figure 4.3:
Scaled lambda ruler
using the **Zoom-in**
command

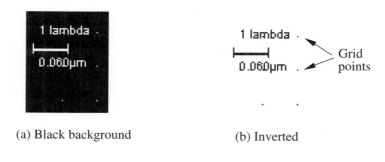

(a) Black background

(b) Inverted

Figure 4.4:
Details of the
grid structure

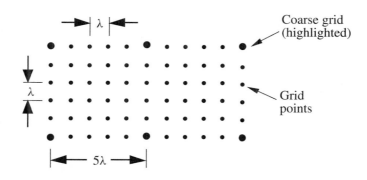

spacing between each grid point is 1 lambda (λ) unit, while the coarse grid spacing is still 5 λ. If you change the zoom setting, the lambda ruler will always show you the scale of the new screen.

The presence of the drawing grid helps the layout designer follow the design rules. It provides a continuous monitor of sizes and spacings and also provides a basis for the relative orientation of pattern features. Chip design requires that we construct lines and boxes with precise dimensions. We will often use the grid and its properties when creating chip layouts.

4.2   Rectangles and Polygons

Now that we have studied the concept of design rules, we can move on to the mechanics of layout. The process itself it straightforward: we create patterns in the work area of the screen using the drawing tools.

Figure 4.5:
Drawing a
rectangle

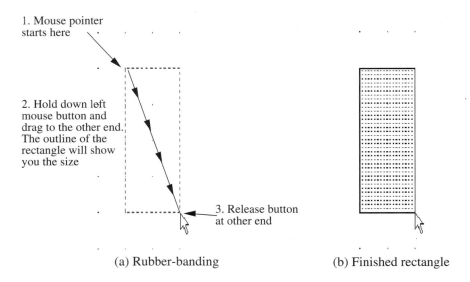

1. Mouse pointer
starts here

2. Hold down left
mouse button and
drag to the other end.
The outline of the
rectangle will show
you the size

3. Release button
at other end

(a) Rubber-banding   (b) Finished rectangle

Most chip designs are based on **Manhattan geometries**, where corners are restricted to 90° (right angle) turns.[1] As mentioned in the previous section, a rectangle is the simplest shape in layout design. The background grid is used to position and size rectangles by incorporating the "snap-to" feature, where corners automatically seek the nearest grid point. The dimensions of a rectangle are restricted to ($n \lambda \times m \lambda$) where n and m are integers. Since the corners are positioned at grid points, the edges (sides) of a rectangle always fall on top a line of grid points.

The easiest way to get familiar with a layout is to actually do it. To this end, let us launch Microwind and direct our attention to leaning how to draw. To create a basic rectangle, we must first put the program into drawing mode by clicking on the on-screen **Draw Box** button. Pushing the button activates the drawing routine for creating rectangles in the work area.

Draw Box
button

To draw a rectangle, first position the on-screen mouse pointer to a spot on the upper-left side, then push and hold the left mouse button down. The location of this point defines one corner of the rectangle. Then drag the mouse (while holding the left button down) towards the lower-right corner. You will see the outline of a changing dashed-outlined rectangle as you move the mouse. Releasing the button defines the opposite corner of the rectangle and it appears on the screen. The procedure is summarized in Figure 4.5(a) and the final (monochrome) result is shown in Figure 4.5(b). The drawing process itself is referred to as "rubber-banding" to describe the behavior of the changing rectangular outline when dragging the mouse.

Chip patterns are created by drawing polygons to represent every feature. We can use rectangles as the primitive object shape to construct an arbitrary (Manhattan) polygon. An

---

1. A major exception to this statement are memory chips, which often permit both 45° and 90° turns to increase the density.

Figure 4.6:
Using rectangles to
build a polygon

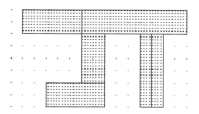 

(a) Rectangle segments   (b) Single polygon

example is shown in Figure 4.6. The screen dumps in Figure 4.6(a) are shown in monochrome mode so that you can see that six individual rectangles have been pieced together to create a polygon. When we change to the color view in Figure 4.6(b), the overall shape appears as a single geometrical object. The apparent boundaries seen in the construction process are not important to either the drawing or the fabrication process. The mask made with this information will show only the single polygon structure. This simple example can be extended to arbitrary shapes, reducing our overall drawing skills down to drawing rectangles and piecing them together. Since a pattern on any given layer is defined by the set of all rectangles, it doesn't matter how the final polygon is constructed.

4.2.1   The Layer Palette Window

Initiating the **Draw** command allows you to construct rectangles in the work area. A layout drawing shows every patterned layer in the CMOS fabrication sequence. Different masked materials are identified using a distinct color-coded fill to allow you view it from the top. If you are in monochrome mode, then the layers are distinguished by the fill pattern.

Palette button

Layers are selected in the Palette Window that appears on the right side of the screen after Microwind has been launched. If it is not visible after some operation, you can always restore it using the **Palette** button shown on the left side of this text. The details of the Palette Window are shown in the screen dump of Figure 4.7. It is your key to creating layer drawings for chip layers, and it also has buttons for layout and simulation macros that are extremely convenient in the design process. Layer selection is achieved using the list that occupies the lower two-thirds of the Palette Window. Once you are in drawing mode, just select a layer and click on it. The rectangles will remain the selected color (layer) until you change the selection or leave the drawing process. You can lock (protect) a layer by clicking on the key symbol to the right of the layer name in the Palette Window.

The layers listed in the Palette Window correspond to the patterned masks that are used in the selected technology. For example, the list in Figure 4.7 shows that there are two polysilicon layers named Polysilicon and Polysilicon2. Not all processes have the second poly option available. Take a moment to familiarize yourself with the layer list. You will see that most of the layers correspond to the masking steps discussed in Chapter 3 for the basic CMOS process flow.

Figure 4.7:
Drawing Palette
Window

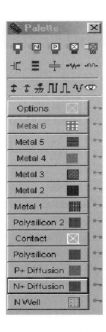

4.2.2   Object Editing

Even the most skilled layout designer will make errors or have to change some features of a drawing. Editing can require that one must move an object, resize it, or even delete it. Microwind provides button commands that initiate these operations in a straightforward manner. Each button is shown at the side of the description below. Simply left-click on the icon to activate it.

It is strongly recommended that you launch Microwind and work along with each example to learn how the editing operations work. This will make it much easier to apply these operations when we start designing MOSFET and CMOS circuits.

Stretch or Move

Stretch, Move
button

This operation allows the designer to change the dimensions of a rectangle by stretching it. You can also move a single rectangle or an entire group of boxes to a different location on the drawing screen.

To stretch a rectangle, activate the command and then point and click on the edge of the rectangle that you want to modify. When you release the button, a dashed outline will appear around the rectangle. It will rubber-band as you move the mouse pointer; note that the cursor will not be touching the rectangle when you stretch it. When the edge is at the desired position, left-click the mouse and a newly sized box will appear. Figure 4.8(a) illustrates the screen view when changing the width of a line, while Figure 4.8(b) shows a stretch in the length.

Figure 4.8:
Stretching a
rectangle

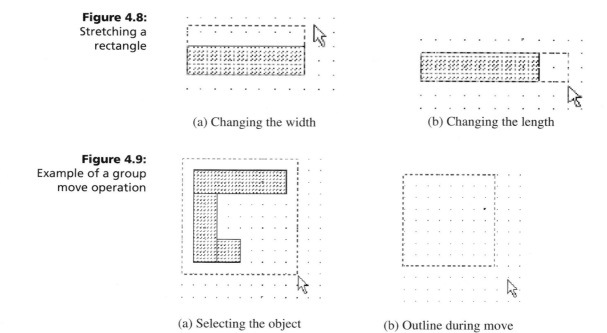

(a) Changing the width   (b) Changing the length

Figure 4.9:
Example of a group
move operation

(a) Selecting the object   (b) Outline during move

You can also use this command to move a single object or a group of objects. First, use the mouse to create a box around the objects you want to move; when moving a group of objects, the relative orientations of selected objects inside the box will remain unchanged. The box is drawn by placing the cursor at a point outside the group, pushing and holding the left button, then drag the mouse until the group is surrounded. An example is shown in Figure 4.9(a). Releasing the mouse button creates a dashed outline of the grouped objects that moves with the cursor. See Figure 4.9(b) for a screen shot during this process. When the outline is in the desired position, left-clicking the mouse leaves the selected group in the new location.

Move Step-by-Step

The move operation is convenient for large translations, but there will be times when we need more precise control. This can be obtained using the Move Step-by-Step operation.

The **Move Step-by-Step** command can be found in the **Edit** submenu, as seen in Figure 4.10; note that the Move command can be also executed from the same submenu. To use this editing function, activate the command, then use the mouse to drag a box around the object(s) that you want to move. When the box is formed and you release the button, an arrow control box will appear with move buttons. This is shown in Figure 4.11. Clicking on the arrows move the selected objects by the number of grid points indicated. The default is 1, but you can adjust it as desired.

The **Move Step-by-Step** command is particularly useful when building up large circuit layouts that must be aligned for connection to, say, a ground connection.

Figure 4.10:
Move Step-by-Step
Command

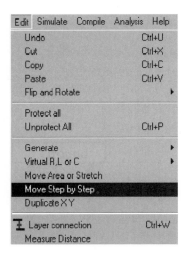

Figure 4.11:
Move buttons

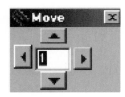

Copy Button

Copy button

The copy operation uses this button to do exactly what it sounds like: it allows you to select a group of objects, copy the image into a clipboard, and then paste the copy somewhere else on the screen. The copy operation does not alter the original objects in any way.

To use this editing procedure, first activate the **copy** button, then draw a box around an object(s) using the same technique as for the move operation. An example is shown in Figure 4.12(a). When you release the button, everything within the box is copied into a clipboard. As you drag the mouse (don't press any buttons yet!), an outline of the selected region moves with the cursor, as in Figure 4.12(b). When you want to copy the pattern

Figure 4.12:
Using the copy
operation

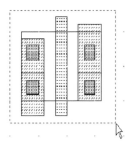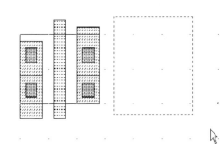

(a) Selecting the object   (b) Outline during copy

Figure 4.13:
Screen shot of
original and copied
patterns

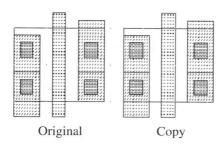

Original   Copy

onto the work area, just left-click the mouse button. The copied object will appear in the location defined by the box at that time. The final result for our example is shown in Figure 4.13.

Many aspects of chip design involve replicating regions of the chip that correspond to a particular operation. The copy operation is extremely useful for reducing design time, and it allows you to reuse any layout that has already been completed.

Delete

Delete button

This is an indispensable operation! It allows you to delete (erase) any object or group of objects in the layout. If you want to delete a single rectangle, just activate the **Delete** button, and then point to the undesired object. A single click will eliminate it entirely. You can also delete a group of rectangles so long as they can be enclosed in a rectangular region with no other objects. Just use the mouse to draw a box around the group. When you release the mouse, all of the patterns within the box will be erased.

Undo

Of course, we do need a safety net just in case we perform an editing operation by mistake. This is the **Undo** command in the **Edit** submenu shown in Figure 4.14. It allows you to

Figure 4.14:
Undo command

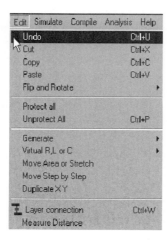

Figure 4.15:
Rotate operation
example

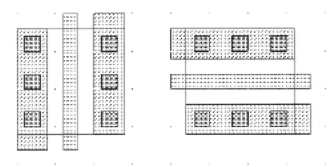

(a) Original orientation   (b) Rotated 90 degrees

go back one operation and returns you to the status of the design before that operation was executed.

The **Edit** submenu listing also shows the keyboard commands for the editing operations:

- Undo **[Ctrl][U]**
- Cut **[Ctrl][X]**
- Copy **[Ctrl][C]**
- Paste **[Ctrl][V]**

These are the same as those used in many popular programs, and tend to become more convenient to use as you become more familiar with the program.

Flip and Rotate

The **Edit** submenu also shows the presence of the flip and rotate operations. These allow you to select a group of objects then change the orientation. A rotate operation is easy to visualize. Since we are working in a Manhattan geometry, Microwind allows 90° rotations either clockwise or counterclockwise as desired. Figure 4.15 illustrates the effect of a single rotate. The original orientation of the pattern is shown in Figure 4.15(a), while Figure 4.15(b) shows the group after a 90° counterclockwise rotation has been completed.

A vertical flip operation turns the object group "upside down" while maintaining the structure of the group; similarly, a horizontal flip operation changes left to right and vice-verse by reflecting the group through an imaginary vertical axis. Figure 4.16 provides an example of a vertical flip. You can see that the patterns have been flipped, but that the overall pattern still looks the same.

Flip and rotate operations are initiated by first using the command sequence

Edit ⇒
    Flip and Rotate ⇒

Figure 4.16:
Vertical flip
example

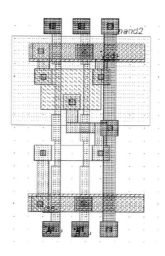 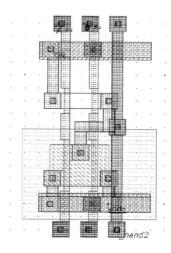

(a) Original pattern   (b) After vertical flip

which places you into a sub-submenu where you can select the desired flip or rotate operation from the list. Then use the cursor to draw a box around the object group that you want to edit. When you release the button, the screen will refresh and give you the newly oriented pattern.

4.2.3   Other Useful Commands

In addition to object editing, Microwind provides a few other useful features that are worth noting.

Protect All/Unprotect All

This feature allows you to "freeze" your designs to guard against accidental editing errors. An unprotected design is one where the characteristics of objects may be changed or erased using the commands discussed above. All layouts are unprotected unless you invoke the **Protect All** command in the **Edit** submenu. A protected layout cannot be altered by any editing actions. Editing privileges are regained by means of the **Unprotect All** command. Microwind identifies protected design areas by changing the fill colors or pattern to transparent objects. A protected rectangle is outlined in the layer color. An unprotected design is shown in Figure 4.17(a); the screen shot in Figure 4.17(b) shows the same area after it has been protected.

Of course, you should save your work continuously as a layout grows larger and more complex.

Figure 4.17:
Using the protect
feature of
Microwind

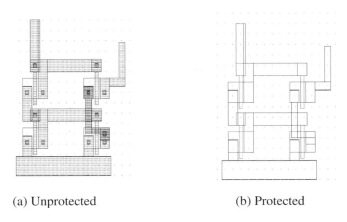

(a) Unprotected   (b) Protected

The command sequence

File ⇒
  Save layout

will save both your design and your emotional stability should something like a power
outage hit. You can also use the keyboard to save by pushing the F2 function key **[F2]**.

Measure Distance

This feature allows you to measure sizes in lambda units. The command sequence is

Edit ⇒
  Measure Distance

**Measure
Distance**
button

You can also use the on-screen distance button shown in the margin. Once executed, place
the mouse pointer at the point where the measure is to begin. Then, push and hold the left
mouse button while dragging the mouse to the other end-point. Releasing the button displays a ruler that is calibrated in units of λ. An example is shown in Figure 4.18. If you

Figure 4.18:
Measuring
dimensions on a
layout drawing

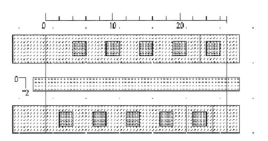

Figure 4.19:
2-dimensional ruler

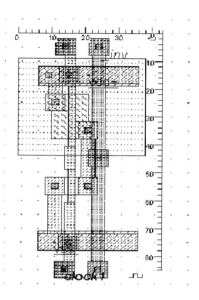

select two points that are not connected by a straight horizontal or vertical line, a 2-dimensional ruler will appear to show the distances. The ruler shown in Figure 4.19 was obtained by defining a line that was started in the upper-left corner of the network and ended at the lower-right corner. Reading values off of the ruler gives a quick estimate for the area of a section in units of λ^2.

The ruler is not a permanent feature of the drawing and will disappear when the next operation is started. Alternately, you can use the command

View $\Rightarrow$

    Refresh

which redraws the layout from the masking information in memory.

Pan and View All

If you are working on a large circuit, you will find it useful to zoom in **[Ctrl][Z]** to work on details. The on-screen arrow buttons on the right side of the Menu bar allow you to "pan" the view, i.e., move around to see the other sections.

Pan button

The operation of the Pan buttons may seem odd until you understand what they represent. The layout drawing itself is assumed to be stationary; in other words, it is like a piece of paper that is taped to a drawing table. The view that you see on the screen corresponds to taking a small picture frame and then looking through the frame as it moves over the surface. This is portrayed in Figure 4.20. The Pan buttons move the location of the frame, not the drawing. Thus, when you push the Up button, the drawing itself will move downward.

Figure 4.20:
Pan operation

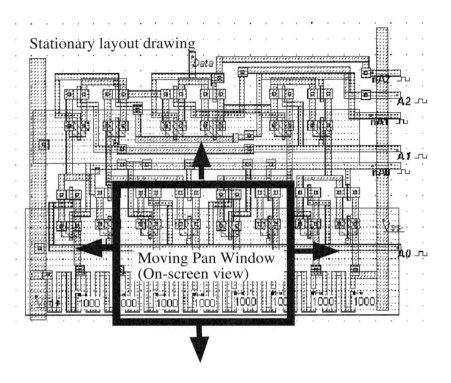

If you are doing detailed work and want to see the entire layout, you can use the command sequence

View $\Rightarrow$
    View All

View All
button

This will automatically zoom out (or in) and show you the entire drawing in a single screen. You may also use the on-screen button shown in the margin or the keyboard combination **[Ctrl][A]** to implement this feature.

The **View** commands are very useful for checking your work visually. Although we have already discussed most of the commands important to layout drawing, it is worthwhile to show a screen shot of the **View** submenu for future reference. Figure 4.21 shows that we have covered all of the commands in the upper half of the window. The lower group are used in the electrical characterization and analysis of the circuits, and will be discussed later. The exception is the Lambda Grid line. When checked (as shown), the grid is displayed automatically in the layout. Selecting it with the cursor hides the grid.

This section has been devoted to describing Microwind's editing commands and showing how they work. You should become familiar with these features by working with the program. Layout design is not difficult to learn, but it can be somewhat tedious until editing shortcuts and tricks become second nature.

Figure 4.21:
The View submenu

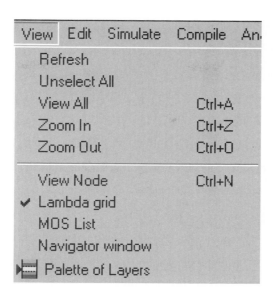

4.3   The MOS Layout Generator Revisited

The MOS Layout Generator, introduced in Chapter 3, provides a simple way to create transistors with specific channel length (*L*) and width (*W*) values. To see this, launch Microwind and push the **FET** button on the upper portion of the drawing Palette Window. This actives the dialog screen shown in Figure 4.22.

Figure 4.22:
The MOS Layout
Generator dialog
screen

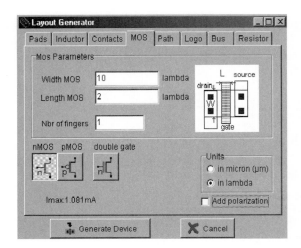

The **Units** in the dialog box are shown set to microns (μm). This automatically interprets width and length values in absolute units. FETs can also be created in lambda measurements by performing a point-and-click operation to the in lambda box. This changes the display to that shown in Figure 4.23. The values of W and L may now be entered in integer multiples of lambda. In the example, $W = 10 \lambda$ and $L = 2 \lambda$. This makes it easier to conform to a set of lambda design rules.

• • • • • • • • • • • • • • •

4.4 Summary

The discussion in this chapter has been directed towards understanding the drawing environment of the Microwind layout editor. It is worth the time to develop your skills by drawing boxes and then testing various commands. At this level, chip design really is a "hands-on" art.

The next chapter presents the entire set of scalable (lambda) design rules that dictate details of mask layout. Examples are provided along the way to keep the presentation from getting too dull. When you finish Chapter 5, you will have the basic knowledge and skills to translate CMOS circuit schematics into layout drawings.

• • • • • • • • • • • • • • •

4.5 References

[4.1]   Uyemura, J. P., *Introduction to VLSI Circuits and Systems*. John Wiley & Sons, New York, 2002.

4.6   Exercises

4.1   Construct a polysilicon box that has dimensions of 2 λ × 12 λ. Use the ruler to measure the size.

4.2   Construct a vertical line on the Metal2 layer with dimensions 3 λ × 14 λ.

4.3   Draw a contact that is exactly 2 λ × 2 λ.

4.4   Construct the polygon with the side lengths shown in the figure on the ndiff layer using rectangles as the primitive element.

Problem 4.4

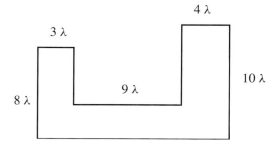

4.5   Construct the polygon with the side lengths shown in the figure on the ndiff layer using rectangles as the primitive element. Then move it four units up.

Problem 4.5

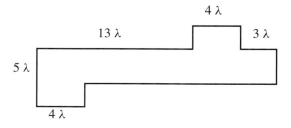

4.6   Draw the polygons shown in the figure using rectangles.

Problem 4.6

4.7 Construct the layout shown in the figure. The ndiff can be drawn as a single rectangle.

Problem 4.7

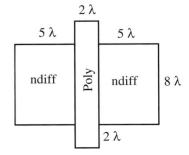

4.8 Construct the layout shown in the figure. Use the widths and spacings shown. Then perform vertical and horizontal flips on the group.

Problem 4.8

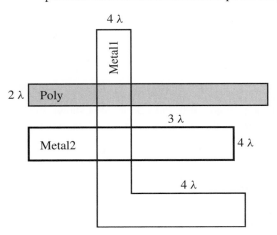

CHAPTER
5

CMOS Design Rules—
Guidelines for Layout

Design rules are a set of specifications that govern the layout of integrated-circuit masking layers. This chapter presents the set of scalable CMOS (SCMOS) rules that are used throughout the book. Scalable rules are also useful for submitting designs to a foundry for fabrication.

5.1 Types of Rules

Chip layout deals with the design of geometrical objects on each masking layer according to a set of rules. In the usage here, a "geometrical object" will imply a polygon that is created by one or more rectangles that are either touching or overlapping. Some examples of simple polygons are shown in Figure 5.1. It is easy to visualize how each can be created using simple rectangles.

Design rules can be classified into four major types:

- **Minimum Feature**—This is the smallest side length of an object on the layer. If the object is a line, then this specifies the **minimum line width**.

- **Minimum Spacing**—The minimum spacing rules govern how close two polygons can be placed.

Figure 5.1:
Examples of
polygons

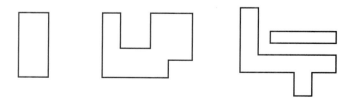

Figure 5.2:
Examples of minimum feature and spacing rules

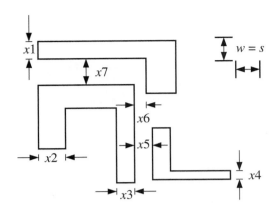

- **Surround**—A surround rule is used when a feature on one layer must be embedded within a polygon on another layer.

- **Exact Size**—An exact size rule means that the feature can only have the dimensions specified in the rule. Other sizes are not permitted.

The minimum width (feature) and spacing are the simplest to visualize. Design rules stipulate minimum values as a limiting factor. We must check to insure that widths and spacings are not smaller than the allowed value. In general, there is no limit to how large the objects are, although other characteristics, such as the electrical response and area consumption, do arise in practice.

Consider the adjacent polygons in Figure 5.2. Several distances labelled $x1, x2, \ldots, x7$ are shown; these are the most obvious dimensions that should be checked. Large side lengths and spacings will not violate any rules, so we will not spend time on them. Suppose that the minimum width and spacing are specified as $w = s$, with the relative size as illustrated in the upper-right side. The critical widths are shown as $x1$, $x2$, $x3$, and $x4$. Of these, $x1$, $x2$, and $x3$ are all $\geq w$. However, $x4 < w$, which is a **design rule violation** that must be changed before we can fabricate the layer. The edge-to-edge spacings are $x5$, $x6$, and $x7$. We see that $x7 > s$, so it satisfies the design rule. However, $x5 < s$ and $x6 < s$, indicating two design violations. To fix these, both spacings must be increased to be $\geq s$. We will rely on an internal design rule check routine to find violations for us.

Minimum width and spacing values are in place to ensure that the structure can be manufactured by the fabrication equipment. Violating the minimum-width rule may result in a "broken" or damaged line segment, leading to a failed device. Placing two features too close together and violating the minimum spacing requirement may result in two incompletely resolved structures, or it may enhance unwanted electrical coupling between the two lines. Either will cause problems in functionality or reliability.

Surround rules originate from the need to embed geometrical features into an existing layer on the chip. An oxide contact cut for access to a pdiff region is shown in Figure 5.3 to illustrate the problem. The cut must be made in an oxide layer that covers a p+

Figure 5.3:
Surround rule
example

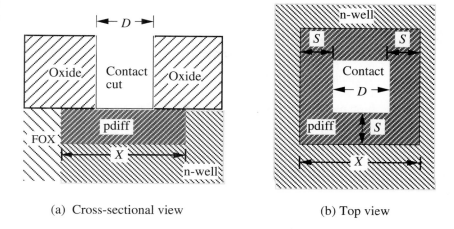

(a) Cross-sectional view   (b) Top view

region that was formed in an earlier step. Figure 5.3(a) shows the cross-sectional view. Since the oxide-cut feature is on a different mask than the pdiff patterns, it must be aligned carefully to ensure the proper placement of the cut over the p$^+$ region. If it touches the n-well, a different electrical connection will be made. Owing to the mechanical positioning of the exposure procedure, we must allow for some mask misalignment to ensure that the contact still falls within the boundaries of the existing layer. The surround spacing, S, in Figure 5.3(b) allows for this to occur. S is chosen to ensure that a reliable contact cut can be made within the worst-case tolerances of the lithographic and fabrication equipment.

Exact size rules usually are associated with contact cuts and vias. These will be seen in the design rule listing.

Now that we have seen the basis for the rules, we note the following:

- Every layer will have minimum width and spacing values for objects on that layer
- Embedded features require surround spacings between two or more masks
- Some design rules specify minimum edge-to-edge spacings between polygons on different layers
- Rules are usually numbered for easy reference. Number conventions vary from line to line, but all are very similar.

In the scalable CMOS rules used here, all spacings are integer multiples of λ, and are called lambda design rules. We will study the rule set for a dual-poly, 6-metal, n-well CMOS processes. The first layer of oxide on the silicon surface (over the poly) will be patterned by the Contact mask. Subsequent metal–oxide layers will be numbered in order of deposition. This means that Metal1 is on top of the Contact oxide, and Oxide1 is on top of Metal1. The via pattern for Oxide1 will be called Via1, and so on.

● ● ● ● ● ● ● ● ● ● ● ● ● ● ●●●

5.2 The SCMOS Design Rule Set

A typical set of SCMOS rules will be listed in the order as they are used in the processing. It is worth repeating that these are **minimum values**. The rule numbers are the same as those used by Microwind. The format will be

> rule# Description: Value

and drawings will be shown for each.

Although we have said that scalable rules can be altered by just changing the value of λ, this is not entirely correct. Over the years, the SCMOS rules have evolved from their original form as CMOS processes changed and spawned new subsets as the line width fell below 1 micron. At the present time, the rule sets break down into roughly three types: generic SCMOS, submicron, and deep submicron. SCMOS can be applied to line widths larger than about 1 micron, while submicron rules typically are used for about 0.8 μm to 0.35 μm. Deep submicron rules are often used below 0.35 μm. The design rule, or DR set, presented in this section are valid for the generic Microwind 1.2 μm process. If you compare the different rule sets, you will see that most values are the same, but some spacings and surrounds may vary by $\pm$ λ. If you are planning to submit a chip design to a foundry, you should check their design-rule set before beginning the layout.

A first reading of a design rule set may seem boring, but you should read through the discussion and examine a few of the rules in detail to get a feeling for what is involved. Familiarity with a DR set comes with usage, so don't try to memorize them. After you tackle some circuit layouts, the rules will seem quite natural and easy to remember.

The first set of rules is shown in Figure 5.4. The N_well mask is used wherever pFETs are to be built. As will be seen later, it is important to provide a positive bias voltage to every n-well region to insure that its voltage is well defined.[1] Wells are quite large because they accommodate pFETs, and a single well can be shared among several pFETs.

The next step in the fabrication sequence is the active area definition which defines the location of every transistor. This is followed by p-type and n-type ion implants that create p^-, p^+ and n^-, n^+ regions, respectively. Although these are separate masking steps, we have combined them together to pdiff and ndiff regions as discussed in the previous chapter. The ndiff and pdiff dimensions are important because they define the channel width W of transistors. The minimum width of ndiff or pdiff regions i s 4λ. This then defines the minimum drawn W value for transistors in the process.

The next set of design rules is shown in Figure 5.5. The first in the list deals with the polysilicon gate, that will be called Poly1 or just Poly. This set of rules is important to FET design, and we will look at them in more detail later in this chapter. The minimum-drawn

---

1. Microwind uses the term "polarization" to refer to bias.

Figure 5.4:
SCMOS design rules
for N_well, ndiff,
and pdiff

N_well

r101   Minimum well size: 12 λ
r102   Well-to-well spacing:11 λ
r103   Minimum surface area: 144 λ^2

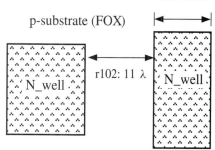

ndiff and pdiff

r201   Minimum n$^+$ and p$^+$ diffusion width: 4 λ
r202   Minimum spacing between two p$^+$ and n$^+$ diffusions: 4 λ
r203   Extension over n-well after p$^+$ diffusion: 6 λ
r204   Minimum spacing between n$^+$ diffusion and n-well: 6 λ
r205   Border of well after n$^+$ bias: 2 λ
r206   Distance between n-well and p$^+$ bias: 6 λ
r210   Minimum surface: 24 λ^2

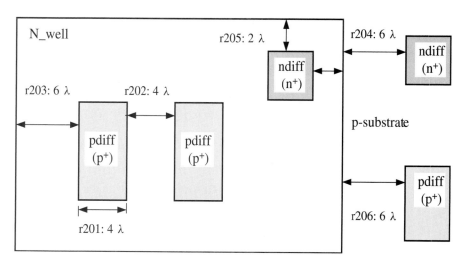

Figure 5.5:
SCMOS design rules
for Poly1 and Poly2

Polysilicon (Poly1)
r301   Polysilicon width: 2 λ
r302   Polysilicon gate on diffusion: 2 λ
r303   Polysilicon gate on diffusion for high voltage FET: 4 λ
r304   Between two polysilicon boxes: 3 λ
r305   Polysilicon versus other diffusion: 2 λ
r306   Diffusion after polysilicon: 4 λ
r307   Extra gate after diffusion: 2 λ

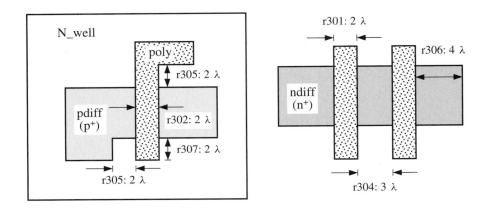

Poly2
r311   Polysilicon2 width: 2 λ
r312   Polysilicon2 gate extended beyond diffusion: 2 λ

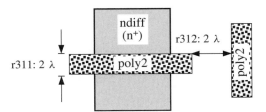

channel length, L, is equal to the smallest-drawn Poly feature size (Rule r301); thus the minimum channel length is $L = 2 λ$. Combining the ndiff/pdiff and poly rules shows that the minimum aspect ratio is

$$\left(\frac{W}{L}\right)_{min} = \frac{4\,λ}{2\,λ} = 2 \tag{5.1}$$

Figure 5.6:
SCMOS design rules
for option layers
and contacts

Option

ropt   2 λ

Border of "option" layer over diff n⁺ and diff p⁺
(In Microwind, this is used to block the etchback
  of the gate oxide)

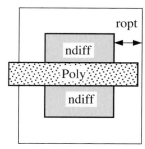

Contacts

r401   Contact size: 2 λ × 2 λ
r402   Spacing between two contacts: 3 λ
r403   Contact to diffusion edge: 2 λ
r404   Poly surround: 2 λ
r405   Metal1 surround: 2 λ
r406   Contact to poly gate: 3 λ

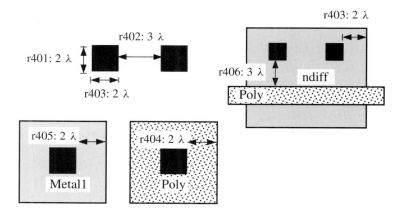

for a transistor with $L = L_{min}$. The gate area consumed by a minimum aspect ratio FET is $(2\lambda \times 4\lambda) = 8\ \lambda^2$.

A second polysilicon layer, Poly2, has been included in the design rules. This is a somewhat specialized layer that is common in many analog CMOS processes where it is used for resistors and capacitors.

The next group of rules is shown in Figure 5.6. The Option layer is a specialized pattern that is used to effect operations, like blocking an ion implant over some areas to increase the resistance. Microwind uses the Option layer to create high-voltage or low-leakage MOSFETs. It can also be specified to block the formation of a silicide contact.

Contact DRs are important considerations in the design. By definition, contacts provide electrical access to ndiff, pdiff, and poly regions by creating cuts in the oxide above these layers. All contacts must be the same size: $2\lambda \times 2\lambda$. This is a fabrication constraint and arises because the processing is optimized to create plugs of a specific size. The edge-to-edge spacing between contacts is 5λ. As we will see later, we usually add as many contacts as can be fit into a specified area to enhance the electrical performance. Contacts are always embedded within features on other layers, so a surround spacing of 2λ is specified. Also note that the spacing between the edge of a poly gate and a contact is 3λ. This will affect the overall layout of FETs.

The next group of design rules deal with the metal and oxide layers. Consider the Metal1 and Via1 rules in Figure 5.7. The minimum width and minimum edge-to-edge spacing for Metal1 patterns are both specified to be 4λ. Via1 cuts are in the oxide above Metal1. The dimensions of Vias are exact with Via1 values of ($2\lambda \times 2\lambda$) and edge-to-edge spacing of 5λ. The rule r603 is the spacing between a Contact and a Via1, and is allowed to be 0 in deep submicron rules. This allows us to stack contacts and vias on top of each other, increasing the integration density over earlier rules that required a non-zero spacing. Vias are embedded objects, so they are subjected to surround spacings; Via1 values are shown to be 2λ. The remaining rules for metals (Metal2 through Metal6) and vias (Via2 through Via6) are similar. These are listed in Figures 5.7 through 5.10. It is worth noting that the minimum-width and spacing values are larger in the highest layers. The values are:

Metal2: 3λ

Metal3: 3λ

Metal4: 3λ

Metal5: 8λ

Metal6: 8λ

The Via5 and Via6 specifications are also larger than lower-level connections.

The final rule listed in Figure 5.10 deals with the design of an input/output (I/O) pad. This is the point where the internal CMOS circuit is interfaced to a pin in the IC package, and routed eventually to the outside world. Pads are specified to be $100\ \mu$m $\times\ 100\ \mu$m to allow a wire to be bonded to it. After working in submicron chip design, $100\ \mu$m may seem huge, but after we think about it, $100\ \mu$m $= 0.1$ mm, which is still quite small by everyday standards! The cut in the passivation layer provides electrical access to the pad, which is designed in the highest metal layer (Metal6 in this rule set). The connection to lower-level metals and the transistors is accomplished using vias and contacts.

Scalable rules are useful in that they can be ported to different processes by changing the value of λ and making minor modifications of some values. It is important to remember that scalable rules do not give the highest packing density. Process-specific values are needed to achieve that goal.

Figure 5.7:
SCMOS design rules
for Metal1, Via1,
and Metal2

Metal1

r501   Metal1 width: 3 λ
r502   Between two Metal1: 4 λ
r510   Minimum surface: 32 λ^2

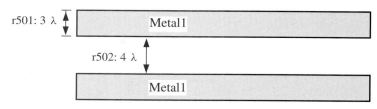

Via1

r601   Via1 size: 2 $\lambda \times 2 \lambda$
r602   Spacing between Via1 edges: 4 λ
r603   Between Via1 and contact: 0 (Can stack in deep submicron rules)
r604   Extra Metal1 over Via1: 2 λ
r605   Extra Metal2 over Via1: 2 λ

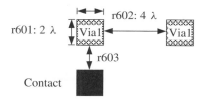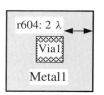

Metal2

r701   Metal2 width: 3 λ
r702   Between two Metal2: 4 λ
r710   Minimum surface: 32 λ^2

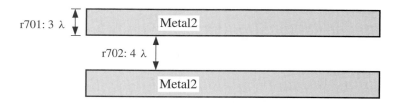

Figure 5.8:
SCMOS design rules
for Via2, Metal3,
and Via3

Via2
r801   Via2 size: 2 λ × 2 λ
r802   Spacing between Via2 edges: 4 λ
r804   Extra Metal2 over Via2: 2 λ
r805   Extra Metal3 over Via2: 2 λ

 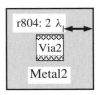 

Metal3
r901   Metal3 width: 3 λ
r902   Between two Metal3: 4 λ
r910   Minimum surface: 32 λ²

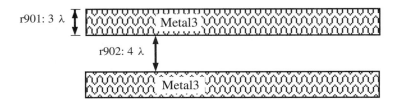

Via3
ra01   Via3 size: 2 λ × 2 λ
ra02   Spacing between Via3 edges: 4 λ
ra04   Extra Metal3 over Via3: 2 λ
ra05   Extra Metal4 over Via3: 2 λ

Figure 5.9:
SCMOS design rules
for Metal4, Via4,
and Metal5

Metal4

rb01   Metal4 width: 3 λ
rb02   Spacing between two Metal4: 4 λ
rb10   Minimum surface: 32 λ²

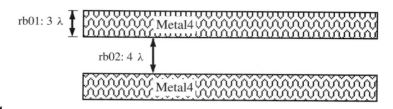

Via4

rc01   Via4 size: 2 λ × 2 λ
rc02   Spacing between Via4 edges: 4 λ
rc04   Extra Metal4 over Via4: 2 λ
rc05   Extra Metal5 over Via4: 2 λ

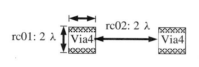

Metal5

rd01   Metal5 width: 8 λ
rd02   Between two Metal5: 8 λ
rd10   Minimum surface: 100 λ²

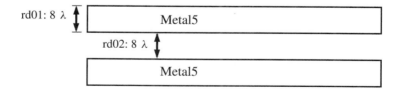

Figure 5.10:
SCMOS design rules
for Via5, Metal6,
and pad design

Via5
re01   Via5 size: 5 $\lambda \times$ 5 λ
re02   Spacing between Via5 edges: 5 λ
re04   Extra Metal5 over Via5: 2 λ
re05   Extra Metal6 over Via5: 2 λ

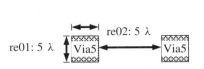

Metal6
rf01   Metal6 width: 8 λ
rf02   Between two Metal6: 15 λ
rf10   Minimum surface : 300 λ^2

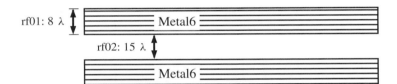

Pad Design
rp01   Pad size: 100 μm $\times$ 100 μm
rp02   Spacing between Pads: 100 μm
rp03   Surround (passivation): 5 μm
rp04   Spacing between Pad and Active: 20 μm

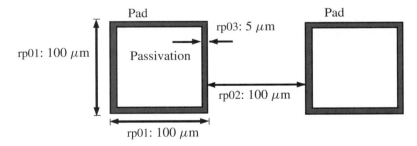

● ● ● ● ● ● ● ● ● ● ● ● ● ● ● ● ●

5.3   FET Layout

The first application of the design rules will be to create nFETs and pFETs with specific W and L values. You will see that the λ values for certain spacings dictate limits for some aspect of the FET geometry. We will detail the procedure first, then see it in action using Microwind. It is important to remember that our discussion employs the SCMOS listing in the previous section. If you use a different set of rules, some of your conclusions may be different.

Let us start with an nFET. When creating layout drawings, you do not need to draw the layers in the order that they occur in the processing. The information on each layer is maintained in an internal database so that you can switch from layer to layer as desired. In this example, we will start with the poly gate since it is the smallest feature in the device. The drawing sequence will be given by:

Poly gate

ndiff

Contact

Metal1

The major steps are shown in Figure 5.11. Consider first the poly-gate pattern. In the finished FET, the gate is a poly line whose width translates into the drawn channel length L. In general, FETs are assumed to all have a value of L equal to the minimum poly line width: 2 λ in design rule r301. This is shown as the first step in Figure 5.11(a).

After the gate is defined, we may add the ndiff region. This is drawn as a single box (not two) since the poly gate will mask the n-type implant when the self-aligned FET is made. This automatically will create separate n[+] regions.[2] The side parallel to the poly gate constitutes the drawn channel width W. The value of W cannot be less than the minimum ndiff feature size in r201 = 4 λ, but adding the contacts to this structure will yield a W larger than this value. If you don't draw the ndiff box correctly the first time, don't

Figure 5.11:
nFET drawing
sequence

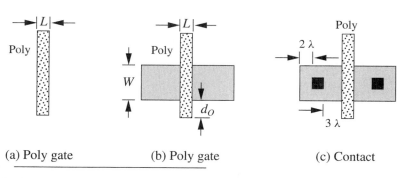

(a) Poly gate   (b) Poly gate   (c) Contact

2.   Remember that poly over ndiff gives an nFET.

Figure 5.12:
Finished nFET

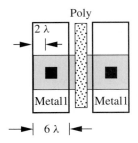

worry, you can always edit it later. Adding the ndiff box gives us the steps illustrated in Figure 5.11(b). Note the extension of the poly gate over the edge of the ndiff; this is called the gate overhang distance, d_o. In terms of the design rule set, we see that rule r307 gives the minimum extension of 2λ.

The next step is to add the contacts so that we can electrically connect ndiff to the rest of the circuit. Since the contact has exact size rules, it has dimensions of $2\lambda \times 2\lambda$. The minimum spacing from the edge of the poly gate to the contact is given by rule r406 = 3λ. Rule r403 gives the minimum spacing of the contact edge to the ndiff edge as 2λ. These are summarized in Figure 5.11(c). With the rules, this design yields a FET with a channel width of

$$W = 2\lambda + 2\lambda + 2\lambda$$
$$= 6\lambda \tag{5.2}$$

because the contact itself is 2λ tall and there are 2λ spacings above and below the contact. The aspect ratio of this transistor is

$$\left(\frac{W}{L}\right) = \frac{6\lambda}{2\lambda} = 3 \tag{5.3}$$

since the gate length was defined to be 2λ.

The finished device is shown in Figure 5.12. We have added metal lines on both sides. The spacing between the contact and Metal1 must satisfy the surround rule r405 of λ on both the left and right sides.

A minimum aspect-ratio device with $(W/L) = 2$ can be drawn to meet the minimum width $r501 = 4\lambda$ design rule, by changing the ndiff region from a simple rectangle to a more complex polygon. The important points in the redesign are to insure that the design rules are satisfied while shrinking the width to 4λ. This includes the contact–metal and contact–ndiff edges, but we also need to look at the poly-ndiff spacing of 2λ given by rule r305. The value of W is defined by the extent of the n^+ region under the polysilicon gate. Reducing the width of the ndiff next to the poly gate and "stretching" the device to satisfy the design rules gives us the layout shown in Figure 5.13. The need for a 2λ spacing between the edge of the poly gate and the ndiff edge results in a transistor of 18λ in the horizontal direction, compared with 16λ in the larger transistor. This example illustrates the fact that the design rules dictate the layout.

Figure 5.13:
Modification to a
minimum-sized FET.

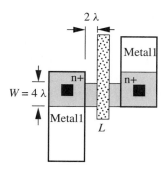

Figure 5.14:   Final
pFET layout

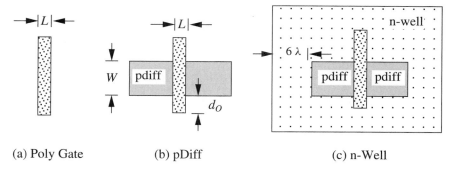

(a) Poly Gate   (b) pDiff   (c) n-Well

A pFET can be constructed in a similar manner. The drawing sequence used here is:

Poly Gate
p-Diff
n-Well
Contact
Metal1

which follows the idea of starting with the gate and diff as the basis for the design. Figure
5.14 shows the sequence. The poly gate is defined in Figure 5.14(a), and the pdiff is added

Figure 5.15:   Final
pFET layout

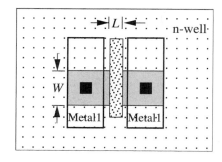

in (b). The rules are the same for the nFET, and the geometry defines the drawn values of *W* and *L*. Adding the n-well gives the layout of Figure 5.14(c). The critical-design rule for the basic structure is the spacing between the pdiff and n-well edges, given here as 6 λ. The remaining steps define the Contact and Metal1 patterns that result in the layout of Figure 5.15.

Example 5.1

Now that we have seen how to draw FETs on "paper" we know how they should look in a layout editor. Launch Microwind and place it in drawing mode by pushing the on-screen button. You will find it useful to use the zoom-in **[Ctrl][Z]** command once to get to the

λ-space grid. The screen shot in Figure 5.16 shows the construction steps.

The drawing Palette menu allows you to select each layer as needed. To draw the transistor gate, click on the Polysilicon bar on the Palette window and then draw a narrow, vertical box. Recall that the minimum width of a poly line is 2 λ. As you draw, a dialog line will appear at the bottom of the work screen that says you are drawing a poly box. As you continue to draw, the dialog line gives the dimensions of the box in both in μm and λ (to the nearest integer value). If you draw a box with a side length that is less than the minimum 2 λ, you will get the warning label **"too small"** on the same line. Microwind automatically detects minimum size violations as you are drawing. When you get to a box around 2 λ wide and 8–10 λ high, release the drawing button. The result will be a red box that snaps to the nearest grid points. The dialog line will inform you that it has stored a poly box with the specified dimensions. If you need to edit the size, either use the **Move/ Edit** button or just add more rectangles. Don't forget the erase button is always available to delete objects.

Figure 5.16: FET drawing in Microwind

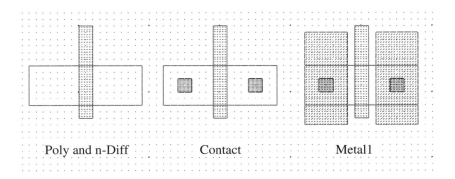

Poly and n-Diff   Contact   Metal1

Figure 5.17:
N-contact button

n-contact button

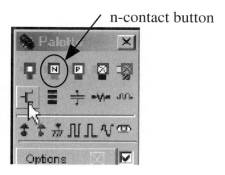

Figure 5.18:
Evolution of the
nFET using a 2-
dimensional
simulation

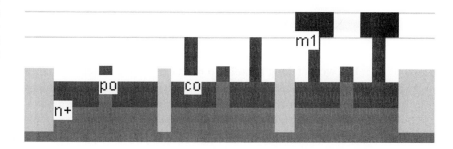

The next step is to draw the ndiff region. Select the n+ Diffusion button from the Palette window (a green box), and draw a rectangle, as shown in Figure 5.16. Try to make the height about 6 λ, and the width greater than about 20 λ (just to be on the safe side). Be sure to let the poly overhang at least 2 λ beyond the ndiff edge, as in the bottom part of the Microwind drawing. Up to this point, your screen should look something like the left-most group in Figure 5.16.

Contacts can be added using the **Contact Layer** button, which is a white box with an "X" in it. You must draw the contact so that it snaps to the exact size of 2 λ × 2 λ because the sizing is not automatic. Alternately, you can select the **n-contact** button on the first row of the Palette window shown in Figure 5.17. This is a macro aid that adds n+, contact, and Metal1 with a single action.

The final step is to add Metal1. Click on the appropriate box in the Palette window and add two vertical metal lines, as in the right-most drawing in Figure 5.16. The Metal1 lines have been drawn 6 λ wide to provide the surround spacing of 2 λ around the contacts. This completes the sequence. To check the result, run a 2-dimensional simulation and select a horizontal line that passes through both contacts. Figure 5.18 shows the evolution of the FET using this technique. Three-dimensional simulation gives the structure shown in Figure 5.19.

Figure 5.19:
3-dimensional view
of the FET design

P Substrate

It is worth taking the time to practice drawing boxes on the grid. Most people find that it is easier to draw the polygons correctly than to go back and edit later. Zooming in so that the single-λ grid points are clearly visible will let you get used to the snap-to feature and also let you see the design-rule spacing close-up. Try to move, copy, and delete polygons using the editing features of the program. They quickly become intuitive and easy to control.

5.3.1 The Design Rule Checker (DRC)

A layout must be free of design rule violations before it goes into the fabrication line. Since checking every polygon for size and spacing is quite tedious, internal routines called **Design Rule Checkers (DRCs)** have been developed to do the work for you. DRCs scan the data file that contains all of the information on the polygon locations and sizes and compare it to a listing of all design rules. Errors are detected and flagged so that the layout designer can correct all violations. In everyday slang, we say that we "DRC the layout." When there are no design rule violations, the layout is "DRC-clean."

5.3.2 Design Rules in Microwind

Microwind adopts a set of design rules that are defined by the **Select Foundry [Ctrl][F]** command. The program loads a text file with a special format that is contained in the Microwind folder; these can be seen by opening the folder and looking for files with names such as *cmos018.rul* and *cmos12.rul*. The .rul extension implies a set of rules, but each file also contains electrical and transistor modeling parameters associated with the process.

A design rule check on a layout drawing is initiated using the **Menu** command sequence:

Analysis ⇒
　Design Rule Checker

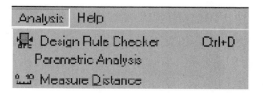

Alternately, the DRC can be started with the keyboard command **[Ctrl][D]** or with the **ON SCREEN** button shown in the margin. Microwind automatically starts checking the layout for design rule violations. If one is found, then a ruler appears with a statement identifying the rule that was not satisfied. As seen in Figure 5.20, the dialog states the problem (spacing between poly is less than 3 lambda) provides the rule number [r301].

Design Rule Checker button

To correct the error, use the **Move/Edit** command to adjust the dimensions as required. Initiating the Edit operation automatically takes you out of the DRC mode. Microwind requires that you fix each problem as it is identified, so it is best to perform a DRC analysis several times as you build the layout. In other layout editors, the DRC will produce a listing of errors referenced to the grid coordinate system that is then used to locate and fix each violation.

The design rules that are loaded into Microwind can be viewed using the command sequence:

Help ⇒
　Design Rules

Figure 5.20: Design rule violation in Microwind

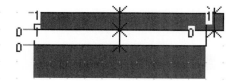

The spacing between poly is less than 3 lambda (r301)

Figure 5.21:
DR List

Layer	Width	Spacing	Surface
	lambda	lambda	lambda2
metal4	4	4	32
via3	2	5	0
metal3	4	4	32
via2	2	5	0
metal2	4	4	32
via	2	5	0
metal	4	4	32
poly	2	3	8
poly2	2	2	8
contact	2	5	0
diffn	4	4	24
diffp	4	4	24
nwell	10	11	144

Design Rules — Design rules and electrical parameters

from the menu bar. This displays a screen, such as that shown in Figure 5.21, that lists the current information. It provides a good online reference for you to access while you are engrossed in layout details.

Microwind can be adapted to any CMOS process by creating a customized .rul file. A sample file for the 0.18 μm technology is listed in Appendix B. The general format for design rules is seen in the poly example:

```
*

*Poly

*

r301 = 2(poly width)
r302 = 2(gate length)
r303 = 4(high-voltage gate length)
r304 = 3(poly spacing)
r305 = 1(spacing poly and unrelated diff)
r306 = 4(width of drain and source diff)
r307 = 2(extra gate poly)

*
```

All numerical values are in units of lambda, and the asterisk (*) denotes a comment line.

To create a custom file, copy one of the .rul files into a new file with a distinct name, such as *mydesign.rul* so that it doesn't overwrite any of the existing files.[3] Then use a text editor to specify values for each listed item. The design rules are numbered following the conventions established in this chapter. For a complete process specification, you will also need to input values for the electrical characteristics, such as MOSFET parameters and parasitic capacitances. These are discussed in more detail in later chapters.

5.3.3   Biasing the Bulk Regions

One rule in standard chip design is that the voltage on every semiconductor n-region or p-region should be well defined for stable operation. The p-substrate of an nFET and the n-well of the pFET are no exceptions, and care must be taken to insure that these are at the proper potentials.

Consider first the p-substrate that acts as the bulk region for an nFET. This should be biased to the lowest voltage in the system. In a single-supply circuit with ground (also called V_{SS}) and V_{DD} connections, this means that p-substrate regions should be connected to ground (V_{SS}). This is achieved by creating a pdiff region in the substrate, adding a contact, and then running a Metal1 line over it that is connected to ground. The pdiff–Metal contact in the Palette window automatically creates the pdiff region, so it is the easiest to use. A substrate ground should be added wherever room permits; a typical rule of thumb is to add a substrate contact every time an nFET is wired to ground.

pFETs are exactly the opposite: the n-well bulk regions must be biased to the highest positive voltage, which is usually V_{DD}. Every n-well is biased in the same manner. An ndiff region in n-well provides a low-resistance region that can be connected to a Metal1 line using a contact. The ndiff–Metal contact generator in the Palette window creates the ndiff region automatically. Biasing the well is very important, so designers often add a V_{DD} to n-well contact every time a transistor is connected to the power supply.

Figure 5.22 shows a cross-sectional view with the p-substrate and n-well bias contacts. The layout example in Figure 5.23 identifies the n-well and p-substrate contacts as related to the V_{DD} and V_{SS} lines. It is important to remember that the substrate and well biases are required by the electrical rules, not the geometrical-layout rule sets. The DRC does not look for the presence of these connections; it only checks the dimensions and spacings of the ones that are present.

---

3. If you do mistakenly overwrite a technology file, it can be restored from the CD.

Figure 5.22:
Substrate and
n-well bias contacts

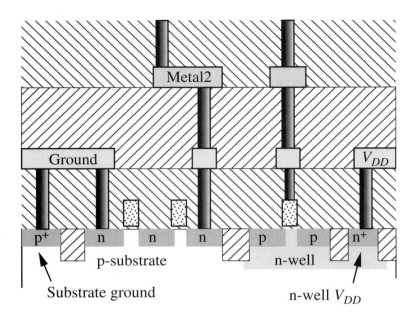

Figure 5.23:
n-well (V_{DD}) and
substrate (V_{SS}) bias
example

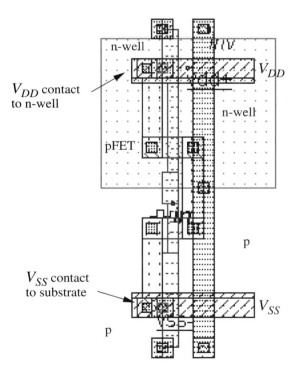

5.4   References

The MOSIS website at **www.mosis.org** has a current listings of the design rules for their processes. The SCMOS (scalable CMOS rules) are lambda based, and the ones of interest will be either the submicron or deep submicron sets. Each process specifies the rule set that should be used. This is probably the best source for up-to-date SCMOS values.

[5.1]   Baker, R. J., Li, H. W., and Boyce, D. E. *CMOS Circuit Design, Layout, and Simulation*. Piscataway, NJ: IEEE Press, 1998.

[5.2]   Clein, D. *CMOS IC Layout*, Boston: Newnes, 2002.

[5.3]   Smith, M. J. S., *Application-Specific Integrated Circuits*. Reading, MA: Addison-Wesley, 1997.

[5.4]   Uyemura, J. P., *Introduction to VLSI Circuits and Systems*. New York: John Wiley & Sons, 2002.

5.5   Exercises

5.1   Draw a poly line that has a width of 1 λ and a length of 4 λ. Draw another poly line that is parallel to it with the same dimensions, and spaced 1 λ. Execute the design rule checker and use it to find and correct the violations.

5.2   What is the minimum spacing between Metal1 and Poly in an nFET that has contacts?

5.3   Suppose that we create two parallel Metal1 lines that are 6 λ wide and 20 λ long, and spaced apart by 6 λ. Is this layout permissible?

5.4   Draw an nFET with $W = 8\ \lambda$ and $L = 2\ \lambda$. Space the ndiff regions to accommodate the contacts to the Metal1 layer. Then run the design rule checker to ensure that you have satisfied the design rules.

5.5   Draw a pFET with $W = 12\ \lambda$ and $L = 2\ \lambda$ within an n-well region. Space the pdiff regions to accommodate the contacts to the Metal1 layer, and also supply a V_{DD} contact to the n-well. DRC the layout to ensure that you have satisfied the design rules.

5.6   Construct an nFET with $W = 10\ \lambda$ and $L = 2\ \lambda$, and contacts to a Metal1 line to provide electrical connections to both ndiff regions. Add a Metal1–Metal2 contact (via) with a minimum-sized Metal2 feature. Then add a Via2 to connect the line to a Metal3 line. This type of stacking is seen in Figure 5.22. Run the design rule checker on your layout.

5.7   Draw a Metal3 line that is 2 λ wide. What indicator on the Microwind screen alerts you to a problem?

MOSFETs— Operation and Analytical Models

MOS field-effect transistors provide all of the switching and amplification functions in CMOS integrated circuits. The speed of a digital chip is directly related to the electrical characteristics of the transistors, which are in turn functions of the layout and processing technology. In this chapter, we examine the basic operation of FETs and how the layout affects the overall performance when used in an IC design.

● ● ● ● ● ● ● ● ● ● ● ● ● ● ●

6.1 MOSFET Operation

There are two levels that can be used to describe the operation of nFETs and pFETs. In the simplest viewpoint, FETs are voltage-controlled switches that are used to steer signals and build logic gates. This is useful for designing logic networks as it concentrates on the switching aspects of transistors. A more accurate approach is to analyze how the current flow is controlled by the applied voltages. Since the switching model is based on the current flow characteristics, we will begin our study of MOSFET modeling at the semiconductor level.

6.1.1 The MOS Capacitor

The MOSFET is built around a small capacitor that consists of a sandwich of three materials: poly-gate, gate-oxide, and semiconductor. As mentioned previously, this is the basis of the acronym MOS from the days when metallic aluminum was used for the gate. The MOS capacitor is integrated into the FET structure so that it can control the charges in the semiconductor.

Figure 6.1 provides a detailed cross-sectional view of an n-channel MOSFET. The MOS layering is identified in the center of the device. A positive gate voltage, V_G, is

Figure 6.1:
nFET cross-sectional
view

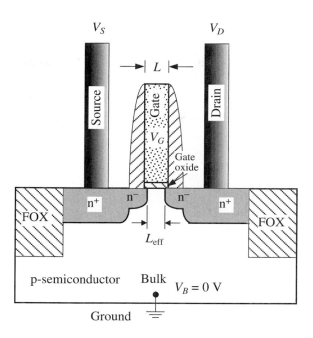

assumed to be applied to the gate from a higher-level metal line (not shown). The p-type substrate is referred to as the **bulk** and the bulk voltage has been set to ground potential: $V_B = 0$ V. The n-type regions in the bulk to the left and right of the gate are electrically connected to a Metal1 line with the Contacts shown. These connections are used to provide the bias voltages, V_D and V_S, to the n-type **drain** and **source** terminals, respectively. The names are determined by the value of V_D and V_S, with the drain being the side at the higher voltage. In most FETs, the drain and source are interchangeable. The drawn channel length, L, is defined by the extent of the poly gate. We have introduced the **effective channel length** L_{eff} that is measured between the edges of the n-type drain and source regions. This distinction is important to the device modeling.

Let us turn to the structure of the MOS capacitor. The gate and bulk regions are electrically separated by the insulating gate-oxide layer, forming the capacitor structure. This region is magnified in the drawing of Figure 6.2. The thickness of the gate oxide is denoted by t_{ox}. If we model this as a simple parallel-plate capacitor, then the gate capacitance, C_G in farads (F), is given by

$$C_G = \frac{\varepsilon_{ins}A}{t_{ox}} \tag{6.1}$$

where A is the area of the plates in cm^2, t_{ox} is in cm, and ε_{ins} is the permittivity of the insulating dielectric in units of F/cm. For silicon dioxide,

$$\varepsilon_{ins} = \varepsilon_{ox} = 3.9\varepsilon_o \tag{6.2}$$

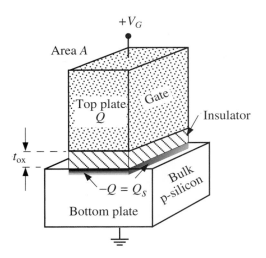

where $\varepsilon_o = 8.854 \times 10^{-14}$ F/cm is the free-space permittivity.

A capacitor is a charge storage device. Applying a voltage V results in a charge of

$$Q = CV \tag{6.3}$$

on the positive plate, and $-Q$ on the negative plate. For the MOS capacitor, the applied voltage is seen to be V_G, since the bulk is grounded. This means that the gate has a positive charge of

$$Q = C_G V_G \tag{6.4}$$

which will be balanced by a charge of $-Q$ in the semiconductor, even though the substrate is a doped p-type. This implies that we can induce negative charge in the bulk by applying a positive charge to the gate. As shown in the drawing, the negative charge is found at the top surface of the silicon (at the silicon-to-oxide interface) and is referred to as the **surface charge**, Q_s. It has two components and is written in general form as

$$Q_s = Q_B + Q_n \tag{6.5}$$

Q_B is called the bulk charge and is associated with the dopants. The remaining term, Q_n, represents a layer of free electrons in the p-type silicon. Bulk charge always exists, but we can control the formation of the electron charge.

A positive gate voltage V_G is needed to form the surface charge, but it must be sufficiently large to induce the formation of the conducting electron charge layer. To create Q_n, V_G must be larger than the (nFET) threshold voltage, V_{Tn}, of the device. This is summarized by the statements

$V_G \leq V_{Tn}$: $Q_n = 0$
$V_G > V_{Tn}$: Q_n layer is formed

The value of V_{Tn} is determined in the fabrication, and is a given parameter to the circuit designer. A typical nominal value is around $V_{Tn} = 0.40$ V. The electron charge can be calculated by modifying the capacitor Q–V relation to read

$$
\begin{aligned}
Q_n &= 0 & &\text{For } (V_G < V_{Tn}) \\
Q_n &= -C_G(V_G - V_{Tn}) & &\text{For } (V_G \geq V_{Tn})
\end{aligned}
\tag{6.6}
$$

This shows that the electron charge builds up linearly with V_G, but it does not start until the gate voltage reaches a value $V_G = V_{Tn}$. With C_G in farads, the electron charge Q_n has units of coulombs (C).

6.1.2   pn Junctions and Channel Formation

The border between an n-type region and a p-type region creates a **pn junction**, which is the most important structure in semiconductor electronics. The existence of pn junctions in MOSFETs and the ability to control the electron charge, Q_n, gives FETs their important switching characteristics.

Let us briefly review the important concepts. Figure 6.3 shows a pn junction and the standard diode circuit symbol. The p-side is called the **anode** and the n-side is called the **cathode**. A diode acts as a rectifying device that allows current flow only in the direction of the arrowhead. In other words, current can flow from anode to cathode, but it is blocked in the opposite direction.

Applying a voltage to the diode gives us two bias conditions, depending upon the polarity. These are shown in Figure 6.4. **Forward bias** is defined as having the positive voltage on the anode and the negative voltage on the cathode and is shown on the right side of the drawing. By convention, the voltage V is positive when applied with this polarity. As seen in the plot, the current flow becomes large when V is a few hundred millivolts positive. The current is given by the familiar relation:

$$
I = I_S(e^{V/V_{\text{th}}} - 1)
\tag{6.7}
$$

Figure 6.3:
Simple pn junction

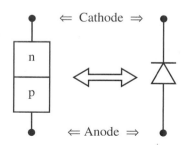

(a) Layering   (b) Symbol

Figure 6.4:
Diode *I–V*
characteristics

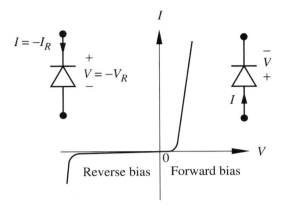

where I_S is the saturation current and V_{th} is the thermal voltage (= 26 mV at 300 K). Switching the polarity gives a state of **reverse bias** which blocks most of the current flow. With the n-side at a higher voltage than the p-side, the current reverses direction as shown:

$$I = -I_R = -I_S \tag{6.8}$$

The reverse current, I_R, increases with the reverse voltage, V_R, but is very small, usually on the order of picoamperes (1 pA = 10^{-12} A). If the reverse voltage is too large, the device undergoes a nondestructive **breakdown**, and the current flows freely in the reverse direction.

Let us now examine the nFET structure redrawn Figure 6.5. Since the substrate is p-type, the n^+/n^- source and drain regions form pn junctions with the bulk. The source voltage V_S can be 0 V or have a positive value, while the drain voltage V_D is larger than V_S by

Figure 6.5:
MOSFET pn
junctions

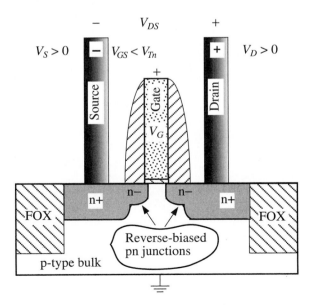

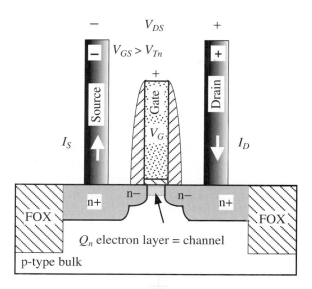

definition. This shows that both the source–bulk and drain–bulk pn junctions are reverse biased, and only small, **reverse-leakage** currents can cross them.

Now consider the drain-to-source voltage:

$$V_{DS} = V_D - V_S > 0 \tag{6.9}$$

The reverse-biased pn junctions between the drain and the source block the current flow between the two terminals. This implies that the drain-to-source path looks like an open circuit, and defines a non-conducting or OFF transistor. Alternately, we can view this as an OPEN switch.

The ability to control the existence of the electron-charge layer with the gate voltage allows us to create a current-flow path between the drain and the source. Figure 6.6 shows the nFET with a gate–source voltage $V_{GS} = V_G - V_S$ that is larger than the threshold voltage:

$$V_{GS} > V_{Tn} \tag{6.10}$$

This creates the electron-charge layer, Q_n, underneath the gate oxide. Since the electrons contact both the n-type source and drain regions, it provides a **channel** for charges to move between them. Applying a drain–source voltage, V_{DS}, gives the currents, I_D and I_S, shown in the contacts. If we ignore leakage currents across the pn junctions, then $I_D = I_S$; for this reason, we usually reference only the drain current.

To summarize our discussion:

$$\begin{aligned}
I_D &= 0 && \text{For } (V_{GS} < V_{Tn}) \\
I_D &\text{ can flow} && \text{For } (V_{GS} \geq V_{Tn}) \text{ and } (V_{DS} > 0)
\end{aligned} \tag{6.11}$$

Figure 6.7:
nFET circuit symbol

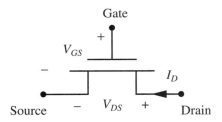

If $I_D = 0$, then the FET is said to be in **cutoff** (or just OFF). When I_D flows, the transistor is said to be ON; alternately, this is sometimes called active operation.

The nFET circuit symbol in Figure 6.7 summarizes the current and voltages in the device. In digital design, the nFET is often visualized as a voltage controlled switch where the gate voltage determines whether the switch is open ($I_D = 0$) or closed (I_D flows).

6.1.3   pFET Operation

A pFET is the electrical complement of an nFET. The cross-sectional drawing in Figure 6.8 shows the details. The source and drain regions are reversed, with the source defined as the terminal having the higher voltage. The control voltages are reversed to read as V_{SG} and V_{SD}, and the pFET threshold voltage, V_{Tp}, is negative by convention: $V_{Tp} < 0$. Typically, $|V_{Tp}| \approx V_{Tn}$, although some processes now allow multiple threshold voltages for both types of FETs.

Figure 6.8:
pFET cross-section

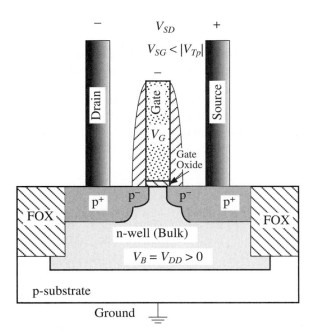

Figure 6.9:
pFET symbol

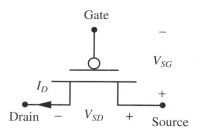

The bulk terminal of the pFET is defined to be the n-well region. To keep the drain–bulk and source–bulk pn junctions in reverse (or zero) bias, we connect the n-well to a positive voltage, shown in the drawing as V_{DD}. In practice, V_{DD} is the positive power-supply voltage used to power the chip, and constitutes the largest DC voltage.

The pFET circuit symbol with the current and voltages identified is shown in Figure 6.9. The bubbled input immediately identifies the polarity as a p-channel device. Although we do not explicitly show the n-well connection, it must always be included in the layout.

The source–gate voltage V_{SG} is used to control the source–drain conduction path. When $V_{SG} \leq |V_{Tp}|$, then no conducting path exists between the source and drain, giving I_D = 0. If we increase the voltage to $V_{SG} > |V_{Tp}|$, then the gate becomes sufficiently negative to induce a positive-charge layer, Q_p, underneath the gate oxide. This is shown in Figure 6.10 and illustrates how the positive-charge layer acts as a channel between the source and

Figure 6.10:
Current flow
in a pFET

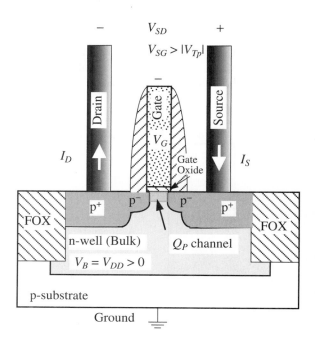

the drain. Current, I_D, flows out of the drain if V_{SD} is applied. pFET operation can be summarized by writing

$$
\begin{aligned}
I_D &= 0 && \text{For } (V_{SG} < |V_{Tp}|) \\
I_D &\text{ can flow} && \text{For } (V_{SG} \geq |V_{Tp}|) \text{ and } (V_{SD} > 0)
\end{aligned}
\tag{6.12}
$$

which is analogous to the nFET. The bottom line is that V_{SG} determines if the device is ON, while V_{SD} provides the electromotive force that moves the charges to give current.

It is important to note that Q_p is a layer of positively charged particles called **holes**. Holes are really "quasi-particles" in quantum theory and represent the absence of electrons in the atomic bonding scheme. Holes are more difficult to move than electrons, which means that nFETs and pFETs will display differences in how they conduct electrical current. In addition to specifying different threshold voltages, we introduce the concept of particle mobility, μ, that has units of cm^2/V-sec. In equal environments, the electron μ_n, is always greater than the hole mobility, μ_p: $\mu_n > \mu_p$. This fact can have a marked effect on how we design CMOS circuits.

6.1.4 Surface Geometry

The current flow through a MOSFET is controlled by the voltages, but the surface geometry plays a major role in how much current the device can conduct. The channel width, W, is the most important dimension.

The top view of an nFET is shown in Figure 6.11. A substrate connection to the ground has been included to show the needed biasing. The drawn nFET aspect ratio, $(W/L)_n$, gives the top-view dimensions of the electron channel formed by Q_n when $V_{GS} > V_{Tn}$. If we increase the channel width, W, then there is a wider area and more charge can flow. Similarly, decreasing W reduces the amount of current that the transistor can handle. This gives the general proportionality rule[1]:

Figure 6.11:
Top view of an nFET

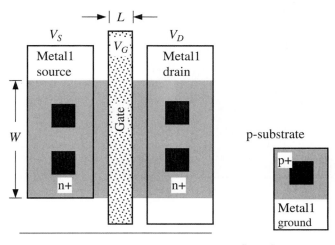

1. In device physics, it is the effective channel length $L_{\text{eff}} < L$ that should be used.

$$I_D \propto \left(\frac{W}{L}\right)_n . \tag{6.13}$$

We can extend this by noting that, since $I = dQ/dt$, the speed of an electron across the channel also affects the current. The electron speed is proportional to the mobility, μ_n. The final dependence to note is that decreasing the oxide thickness, t_{ox}, increases the capacitance, indicating that more charge can be induced for a given voltage. Since W and L are variables of the layout, we chose not to use the total gate capacitance, C_G, in Equation (6.1) since the area, $A = WL$, will change from device to device. It is simpler to introduce the (gate) **oxide capacitance** per unit area as

$$C_{ox} = \frac{\varepsilon_{ins}}{t_{ox}} \tag{6.14}$$

that has units of F/cm^2. Combining these factors leads to the general dependence,

$$I_D \propto (\mu_n C_{ox})\left(\frac{W}{L}\right)_n \tag{6.15}$$

or

$$I_D \propto \beta_n \tag{6.16}$$

where

$$\beta_n = k_n' \left(\frac{W}{L}\right)_n \tag{6.17}$$

is called the **device transconductance** and has units of A/V^2. In this expression we have introduced

$$k_n' = \mu_n C_{ox} \tag{6.18}$$

This is called the **process transconductance** since it is determined solely by processing parameters; it also has units of A/V^2.

The layout considerations for a pFET are similar. Refer to Figure 6.12, where an n-well contact to the **power-supply voltage**, V_{DD}, has been included. The only major difference is the change in polarities of the silicon regions. We can thus conclude that the current flow has the dependence

$$I_D \propto \beta_p \tag{6.19}$$

Figure 6.12:
pFET layout
geometry

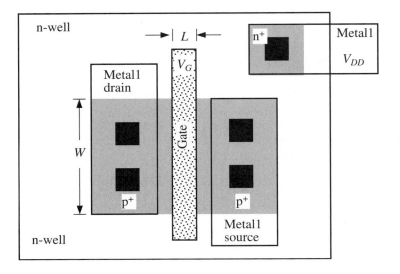

where

$$k_p' = \mu_p C_{\text{ox}}$$

$$\beta_p = k_p' \left(\frac{W}{L}\right)_p \tag{6.20}$$

define the pFET process and device transconductance, respectively. Both quantities have units of A/V^2 since the aspect ratio is unitless.

One difference that is observed at the device level is that the electron and hole mobilities are different. Taking the ratio of process transconductance equations gives

$$\frac{k_n'}{k_p'} = \frac{\mu_n C_{\text{ox}}}{\mu_p C_{\text{ox}}} = \frac{\mu_n}{\mu_p} = r \tag{6.21}$$

where r is the **mobility ratio**. The exact value of r varies with the processing, as many factors affect the mobility. However, $r > 1$ is almost always a valid assumption, with typical values of r ranging from three to perhaps eight or more. Note that the ratio of device transconductances for two FETs of opposite polarity is

$$\frac{\beta_n}{\beta_p} = \frac{k_n'\left(\frac{W}{L}\right)_n}{k_p'\left(\frac{W}{L}\right)_p} = r\frac{\left(\frac{W}{L}\right)_n}{\left(\frac{W}{L}\right)_p} \tag{6.22}$$

This can be adjusted by varying the channel width values, W_n or W_p.

This short analysis demonstrates the interplay between the layout geometry and the current flow levels. Much CMOS circuit design is concerned with selecting aspect ratios,

(*W/L*), for each transistor to obtain the current levels needed for the function at hand. We will examine this dependency in both digital and analog circuits in later chapters.

6.2   MOSFET Switch Models

Digital CMOS circuits are easy to construct using simple switch models to describe nFETs and pFETs. If you have a background in basic digital logic, then learning them is straightforward. And, once mastered, you will gain quick entry into the field of digital chip design.

Binary digits (**bits**) *a*, *b*, *c*, … are unique in that they can only have values of 0 or 1. In a digital integrated circuit, we use time-varying voltages, *V*, at various points to represent the value of a logical bit. Electronic chips obtain energy from an externally supplied power-supply voltage, V_{DD}, as illustrated in Figure 6.13. The value of V_{DD} varies with technology. Classically, $V_{DD} = 5$ V was used to provide an easy interface with bipolar TTL chips. Modern CMOS processes use reduced power-supply voltages to handle power-dissipation and oxide-breakdown problems. Values as low as 1 or 2 V are now commonplace. We do assume that V_{DD} is larger than V_{Tn} and $|V_{Tp}|$ for normal CMOS operation. Regardless of the actual value of V_{DD}, it can be partitioned into "low" and "high" ranges that represent 0 and 1 binary levels. The voltages in between the 0 and 1 ranges are shown with a question mark (?), as they are classified as undefined in the Boolean sense. Note that we have included the designation of the ground as a power supply of $V_{SS} = 0$ V in the drawing. This is a leftover from the days when we used two (or more) power supplies, but it is still used.

The translation between binary variables and voltages allows us to construct very simple models that view FETs as logic-controlled switches. Consider the nFET in Figure 6.14(a). The gate has a logical input *A* that can be a 0 or a 1, corresponding to a low voltage or a high voltage, respectively. Applying a value of *A* = 0, as in Figure 6.14(b), is equivalent to having a low gate voltage. This is not sufficient to create the electron channel, so the nFET is in cutoff with $I_D = 0$. In the switch-level model, this is represented by an open circuit. If *A* = 1, then a large gate voltage has been applied to the device. This induces the formation the **electron charge layer**, Q_n, underneath the gate oxide. Current, I_D, can flow between the drain and the source, and the transistor acts like a closed switch.

Figure 6.13:
Logic voltage levels

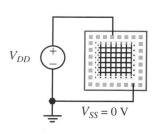
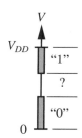

Figure 6.14:
nFET switching
model

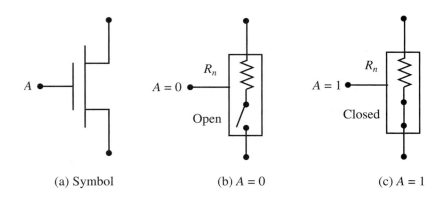

(a) Symbol   (b) $A = 0$ (c) $A = 1$

This is shown in Figure 6.14(c). The nFET **drain-source resistance**, R_n, has been included to model some of the electrical effects. The switching box is viewed as a **logic-controlled element** whose state (open or closed) is implied by the control variable A.

The behavior of a pFET switch is exactly the opposite. In general, pFETs are referenced to the positive power supply since V_{DD} is tied to the n-well. The general symbol is shown in Figure 6.15(a). The input bubble should be interpreted as a logic-inversion symbol. When we apply $A = 0$ [see Figure 6.15(b)], the inversion gives an effective "1" on the switch input. This results in a closed switch in which current, I_D, can flow between the top and bottom terminals. Physically, the conduction layer of positive charge, Q_p, exists because the gate is more negative than the bulk (n-well). R_p represents the pFET drain-to-source resistance. When $A = 1$, the inversion bubble flips the bit so that the switch itself sees a "0" input, and gives an open circuit.

Recall that the C in CMOS stands for complementary and describes a circuit-design technique. In standard CMOS logic, nFETs and pFETs are used in pairs with a common gate input. A general **complementary pair** is shown in Figure 6.16(a). Since the nFET (Mn) and the pFET (Mp) have opposite switching characteristics, using a common input A for both means that only one of the FETs is ON (a closed switch) at a time. This can be verified by example. Figure 6.16(b) shows the case when $A = 0$; in this case, the pFET is ON and the nFET is OFF. Conversely, $A = 1$ forces the pFET into an OFF state while the

Figure 6.15:
pFET switching
models

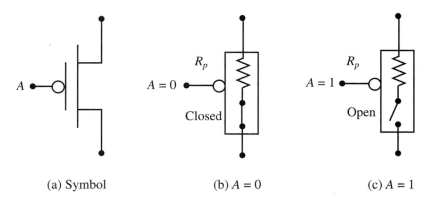

(a) Symbol   (b) $A = 0$ (c) $A = 1$

Figure 6.16:
CMOS
complementary
pair

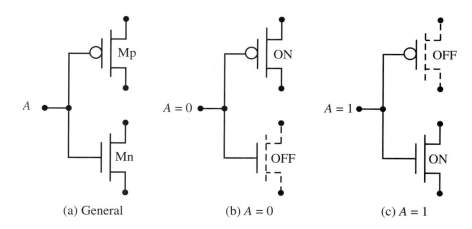

(a) General　　　　　(b) $A = 0$　　　　　(c) $A = 1$

nFET is ON; this is shown in Figure 6.16(c). The operations of the two FETs thus complement each other, giving rise to the name.

Switch models are extremely useful for developing logic gates in CMOS. They are intuitive and simple, which makes them attractive for describing complex switching networks. They do have the drawback that they do not contain sufficient information to calculate switching speeds and other important chip characteristics. And neither FET can pass the full range of voltages on a chip, so that we must exercise care when using the switch equivalents. Electrical modeling is required to round out the design cycle.

● ● ● ● ● ● ● ● ● ● ● ● ● ● ● ● ●

6.3　The Square Law Model

Analytical models are useful because they allow us to compute currents from device voltages using closed-form equations. The **square law model** is the simplest analytical description of a MOSFET, and it is used for design estimates and basic calculations.

Consider the nFET symbol in Figure 6.17. The objective of analytical device modeling is to obtain equations for the drain current, I_D, in terms of the voltages, V_{GS}, and V_{DS}. The square law model has three distinct regions of operation, depending upon the voltages.

Region I: Cutoff

Cutoff occurs when $V_{GS} \leq V_{Tn}$ and is described by

$$I_D = 0 \tag{6.23}$$

The next two operational modes occur when the device is ON with $V_{GS} > V_{Tn}$. The dividing line is the **saturation voltage**,

$$V_{\text{sat}} = V_{GS} - V_{Tn} \tag{6.24}$$

Figure 6.17:
nFET currents
and voltages

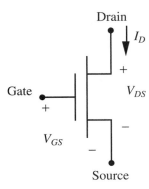

which is a special value of the drain–source voltage, V_{DS}, that is established by the value of V_{GS}.

Region II: Triode Region

This is defined as having $V_{GS} > V_{Tn}$ and $V_{DS} < V_{sat}$, i.e., a small drain–source voltage. The current–voltage dependence for this case is

$$I_D = \frac{\beta_n}{2}[2(V_{GS} - V_{Tn})V_{DS} - V_{DS}^2]$$ (6.25)

This is also called the **linear region** because it is linear in V_{GS}, or the **non-saturation region**, since the drain–source voltage generally is less than V_{sat}. The saturation point is reached when $V_{DS} = V_{sat}$. Substituting gives the saturation-point current as

$$I_{D,\,sat} = \frac{\beta_n}{2}(V_{GS} - V_{Tn})^2$$ (6.26)

Region III: Saturation

The saturation region is where $V_{GS} > V_{Tn}$ and $V_{DS} \geq V_{sat}$. The current-voltage dependence is written by modifying $I_{D,sat}$ with a function of V_{DS} in the form:

$$I_D = \frac{\beta_n}{2}(V_{GS} - V_{Tn})^2[1 + \lambda(V_{DS} - V_{sat})]$$ (6.27)

The parameter λ is called the channel-length modulation factor and has units of $1/V$. For a given value of V_{GS}, the drain current increases linearly with V_{DS}.

The characteristics of the square law model usually are displayed with two plots. The first is a plot of I_D versus V_{DS} for different values of V_{GS}. Varying the gate–source voltage gives the family of curves shown in Figure 6.18. The parabola separating the triode and the saturation regions is the saturation point, where $V_{DS} = V_{sat} = V_{GS} - V_{Tn}$. Note that cutoff is represented by the horizontal axis where $I_D = 0$. The tilt of the curves in saturation is due to the channel-length modulation factor. Another way to view the characteristics is to bias

Figure 6.18:
nFET family
of curves

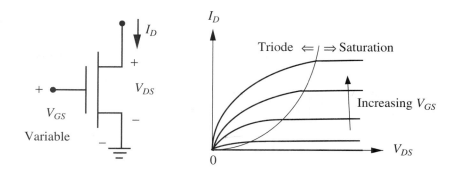

the transistor into saturation with $V_{DS} > V_{sat}$ and plot I_D as a function of V_{GS}. This is shown in Figure 6.19 and clearly illustrates the turn-on threshold effect at a gate–source voltage of $V_{GS} = V_{Tn}$.

Figure 6.19:
Saturation
characteristics

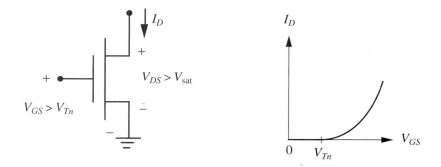

6.3.1   pFET Equations

The square law model may also be applied to describe the pFET by making the appropriate changes. Figure 6.20 shows the device voltages for this case.

Cutoff occurs when $V_{SG} \leq |V_{Tp}|$, which gives $I_D = 0$. Increasing the source-gate voltage to $V_{SG} \geq |V_{Tp}|$ forms a positive-charge channel and allows current to flow between the source and drain. The pFET saturation voltage is defined by

$$V_{sat} = V_{SG} - |V_{Tp}| \tag{6.28}$$

The pFET is in the triode region when the source-drain voltage satisfies $V_{SD} \leq V_{sat}$. The current is given by

$$I_D = \frac{\beta_p}{2}[2(V_{SG} - |V_{Tp}|)V_{SD} - V_{SD}^2] \tag{6.29}$$

Figure 6.20:
pFET voltages

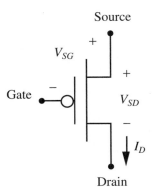

The saturation region operation occurs with $V_{SD} \geq V_{\text{sat}}$, and the current is described by

$$I_D = \frac{\beta_p}{2}(V_{SG} - |V_{Tp}|)^2\left[1 + \lambda(V_{SD} - V_{\text{sat}})\right] \qquad (6.30)$$

The general features of the plots are the same as for the nFET.

6.3.2   Body Bias

Recall that the p-substrate bulk terminal of nFETs is assumed to be grounded. Holding this at a constant voltage introduces a secondary effect that changes the threshold voltage of the transistor. Defining a body bias voltage as

$$V_{SB} = V_S - V_B \qquad (6.31)$$

as shown in Figure 6.21(a), we find that the threshold voltage of the transistor varies according to

Figure 6.21:
Body bias
in an nFET

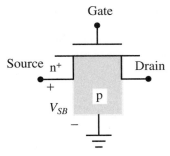

(a) Body-bias voltage

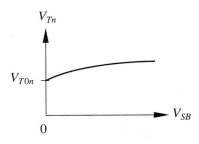

(b) Threshold voltage variation

$$V_{Tn} = V_{T0n} + \gamma(\sqrt{2|\phi| + V_{SB}} - \sqrt{2|\phi|}) \tag{6.32}$$

where V_{T0n} is the zero **body-bias threshold voltage**, γ is called the **body-bias coefficient** with units of $V^{1/2}$, and $|\phi|$ is called the bulk Fermi potential with units of V. Figure 6.21(b) shows how the threshold voltage of the nFET increases with V_{SB}. By convention, we always quote the value of V_{T0n} as "the" threshold voltage in a process specification. Body-bias effects may be small in modern technology because γ is proportional to t_{ox}.

A pFET also has body-bias effects, except that the polarity is reversed since the n-well is biased to V_{DD}. Denoting the zero body-bias threshold voltage by V_{T0p}, the threshold voltage is a function of the pFET body bias, V_{BS}, as described by

$$V_{Tp} = V_{T0p} - \gamma(\sqrt{2|\phi| + (V_{DD} - V_{BS})} - \sqrt{2|\phi|}) \tag{6.33}$$

where the numerical values for γ and ϕ are different for nFETs and pFETs. This says that increasing the body-bias voltage V_{BS} makes V_{Tp} more negative; alternately, we say that increasing V_{BS} increases $|V_{Tp}|$.

6.3.3 Limitations

The square law model provides simple, closed-form expressions for the drain current as a function of the device voltages. However, a closer examination reveals that it is only valid for **long-channel** devices (i.e., those with channel lengths of $L > 10$ μm), as it ignores many effects that arise as the dimensions are reduced. The simple expressions tend to overestimate currents by a large amount and do not give the correct voltage dependences. Short-channel and submicron phenomena modify the charge transport, which change both the shape of the curves and the values. Analytical models have been developed, but they are much more complicated. One simple modification is to change the saturation current to

$$I_{D, \text{sat}} \approx \frac{\beta_n}{2}(V_{GS} - V_{Tn})^{\alpha} \tag{6.34}$$

where α varies with the technology. Currently, $\alpha \approx 1.2$ to 1.5 for submicron devices. More complex formulations change the form of the equations entirely.

These statements should make you wonder why we took the time to study the square law equations in the first place. First, real FETs have I–V curves that *resemble* those presented in this section. Second, square law equations are still useful for doing simple analyses of circuits that show us the general behavior of a network. We don't expect the numbers to be precise, but the important dependences will still come out of the calculations. This allows us to do initial design estimates. Finally, we will always turn to computer simulations to obtain more accurate values. This allows us to see the performance of the circuit with the improved modeling and make design changes if necessary.

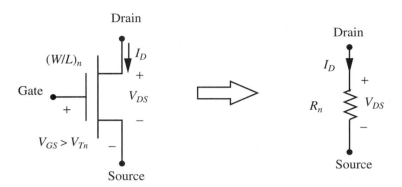

Figure 6.22:
The drain–source
resistance in an
nFET

● ● ● ● ● ● ● ● ● ● ● ● ● ● ● ● ●

6.4   MOSFET Parasitics

Although MOSFETs are used primarily for switching and gain, they always introduce parasitic resistance and capacitance that affects the circuit operation. It is important to understand the basis of these elements, as they cannot be ignored in high-speed chip design.

6.4.1   MOSFET Resistance

MOSFETs are nonlinear devices in that I is not proportional to V. This is seen from the square law models. Resistance is usually implied to be a linear quantity such that **Ohm's Law**, $V = IR$, is valid. Despite this fact, it is often useful to introduce **linear time-invariant (LTI)** resistances that provide some information about the drain–source current flow. The general nFET model is shown in Figure 6.22, where a FET that is biased into conduction is to be represented by anLTI resistor, R_n. This is the same model that was introduced in the switch model.

Figure 6.23:
Determining FET
resistance

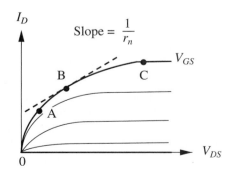

Since the FET is intrinsically nonlinear, it is not clear what value of resistance would be a good choice. Figure 6.23 illustrates the problem. Three points labeled A, B, and C are indicated on the I_D versus V_{DS} curve. The resistance at any point is

$$r_n = \frac{\partial V_{DS}}{\partial I_D} \tag{6.35}$$

so that the slope is equal to $(1/r_n)$. However, since the curve is not a straight line, the value of r_n depends upon the actual point where the slope is taken. This shows that r_n is not a constant, but changes with the voltages; this is an example of a **nonlinear resistance**.

There is no universal choice for selecting a value for the LTI resistor, R_n. However, since I_D is proportional to β_n, most models agree with the inverse proportionality:

$$R_n \propto \frac{1}{\beta_n} = \frac{1}{k'_n(W/L)_n} \tag{6.36}$$

Since β_n has units of A/V^2, we need to multiply the denominator by a factor that has units of volts to obtain the proper units of resistance (V/A). The simplest expression to remember is obtained by using a device at the saturation point with drain–source voltages of $V_{DS} = V_{GS} - V_{Tn}$ and $V_{GS} = V_{DD}$ applied. This gives

$$R_n = \frac{(V_{DD} - V_{Tn})}{I_{D,\,sat}} = \frac{2}{\beta_n(V_{DD} - V_{Tn})} \tag{6.37}$$

Alternately, if $V_{DS} = V_{DD}$ and $V_{GS} = V_{DD}$, we have

$$R_n = \frac{2V_{DD}}{\beta_n(V_{DD} - V_{Tn})^2} \tag{6.38}$$

which is slightly larger than the simple expression. This can be modified to an α-power estimate by writing

$$R_n = \frac{2V_{DD}}{\beta_n(V_{DD} - V_{Tn})^\alpha} \tag{6.39}$$

which is larger yet.

More complex equations have also been proposed. For example,

$$R_n = \frac{V_{DD}}{\beta_n(V_{DD} - V_{Tn})^\alpha} + \frac{V_{DD}}{\beta_n\left[(V_{DD} - V_{Tn})V_{DD} - \dfrac{V_{DD}^2}{2}\right]} \tag{6.40}$$

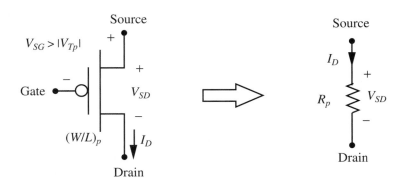

adds terms for both saturation and non-saturation regions with $\alpha = 1.3$ for a short-channel device. This is obviously larger than the other estimates, but has been shown to work well in certain types of gate arrays.

Which expression should you use? Although there may be those that have a preference, none are really correct over the entire range of operating voltages, since a FET is in fact a nonlinear device: it is not possible to define a linear resistance precisely. This, however, is not really important, so long as you remember that this is only a first-order model, not an accurate value. The important dependence on *W/L* is still there, allowing us to use it for first-estimate designs.

The pFET resistance, R_p, can be obtained using the same type of reasoning. For the device in Figure 6.24, we can estimate

$$R_p = \frac{2}{\beta_p(V_{DD} - |V_{Tp}|)} \tag{6.41}$$

for the simplest model, or

$$R_p = \frac{2V_{DD}}{\beta_p(V_{DD} - |V_{Tp}|)^\alpha} \tag{6.42}$$

with $\alpha = 2$ for square law approximations. The expanded equation is

$$R_p = \frac{V_{DD}}{\beta_p(V_{DD} - |V_{Tp}|)^\alpha} + \frac{V_{DD}}{\beta_p\left[2(V_{DD} - |V_{Tp}|)V_{DD} - \dfrac{V_{DD}^2}{2}\right]} \tag{6.43}$$

Regardless of the model you use, you should be consistent between nFETs and pFETs. In other words, once you have decided on an expression for one type of FET, use the same form for the opposite polarity device.

Figure 6.25:
MOSFET
capacitances

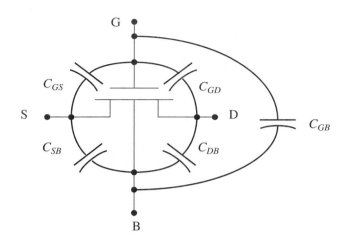

6.4.2   Capacitances

All CMOS circuits have capacitive nodes. This is because of the fact that the MOSFET itself is designed with a capacitor to control the current flow. Unfortunately, capacitances tend to slow down the switching speed of circuits, because it takes a finite amount of time to charge and discharge them. This can be seen in capacitor *I–V* relation of

$$i = C\,\frac{dV}{dt} \approx C\,\frac{\Delta V}{\Delta t} \qquad (6.44)$$

which says that it is not possible to change the voltage instantaneously (with $\Delta t \to 0$), since that would require an infinite current.

MOSFET capacitors are defined between pairs of terminals, including the bulk (p-substrate or n-well) connections. This gives rise to five contributions shown in Figure 6.25. Three of these are related to the MOS capacitor: C_{GS}, C_{GD}, and C_{GB}. The remaining two (C_{SB} and C_{DB}) are due to the pn junctions. All five are nonlinear in that they are functions of the voltages. For analytical estimates, we usually construct LTI values that are reasonable for first designs. Including the nonlinear features requires the use of a simulation program.

The MOS capacitors can be expressed in terms of the gate capacitance,

$$C_G = C_{\text{ox}}WL \qquad (6.45)$$

where the area has been taken to be $A = WL$, as seen in the drawing of Figure 6.26. The coupling between the gate and the source, the drain, or the bulk depends on the operational mode. In cutoff, the dominant contribution is $C_{GB} \approx C_G$, while C_{GS} and C_{GD} are approximately 0 because no channel exists to couple the source or drain to the gate. If the channel

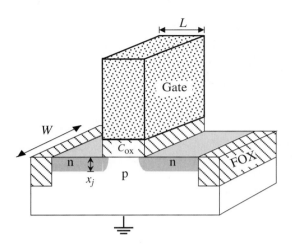

is formed with $V_{GS} > V_{Tn}$, then we have that $C_{GB} \approx 0$, since the electron layer shields the bulk from the gate fields. In saturation,

$$C_{GS} \approx \frac{2}{3}C_G \qquad C_{GD} \approx 0 \tag{6.46}$$

while the triode (non-saturation) region gives

$$C_{GS} \approx \frac{1}{2}C_G \qquad C_{GD} \approx \frac{1}{2}C_G \tag{6.47}$$

Since it can be difficult to track the changes when doing hand calculations, we often use the triode values for LTI design estimates.

The drain–bulk and source–bulk capacitances originate from the reverse biased pn junction. Figure 6.27 shows the geometric details. A pn junction exists at the bottom of the n-region and also along the channel sidewall, as shown in the drawing. The remaining three sidewalls are bordered by field oxide (FOX) and do not contribute to pn-junction capacitance. The total capacitance of an n-type source or drain is written as the sum of bottom and sidewall contributions in the form

$$C_n = C_{\text{bot}} + C_{sw}. \tag{6.48}$$

Device physics provides us with a value of the junction capacitance per unit area, C_j in F/cm^2, so that

$$C_{\text{bot}} = C_j XW \tag{6.49}$$

Figure 6.27:
Junction
capacitance
calculation

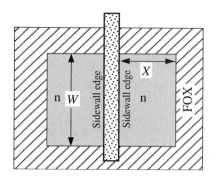

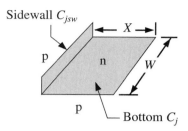

For the sidewall contribution, we define the sidewall capacitance per unit length, $C_{jsw} = C_j x_j$, in F/cm, where x_j is the depth of the pn junction. The sidewall contribution is then

$$C_{sw} = C_{jsw}W \tag{6.50}$$

Summing gives

$$C_n = (C_j X + C_{jsw})W \tag{6.51}$$

which can be computed from the layout geometry. In general, we write

$$C_n = C_j A_n + C_{jsw}P \tag{6.52}$$

where A cm^2 is the **source or drain area** (A_S or A_D, respectively), and P in cm is the **perimeter length** around the source (P_S) or drain (P_D), where there is a sidewall junction. The pn-junction capacitance is nonlinear, and this represents the value with an applied voltage of $V = 0$ V. C_n decreases with increasing reverse-bias voltage, V_R, so we will often use the zero-bias value since it represents the maximum.

It is worth mentioning that all of the FET capacitances are proportional to the channel width, W. This is in contrast to the FET resistance, which is inversely proportional to W. The product RC is a time constant, τ, with units of seconds, so that quantities such as

$$\tau = R_n C_n \tag{6.53}$$

are independent of W. When applied to a technology, this value gives an indication as to the maximum digital-switching speed that is possible.

● ● ● ● ● ● ● ● ● ● ● ● ● ● ●

6.5   Comments on Device Layout

The discussion in this chapter illustrates that the electrical characteristics of a MOSFET are determined by the layout geometry. The channel length, L, is usually the smallest poly line width, so that the channel width, W, becomes the key design variable.

Selecting W determines (along with the processing parameters)

- The device transconductances, β_n, and β_p (proportional to W)
- The transistor linearized resistances, R_n and R_p (inversely proportional to W)
- All parasitic capacitances, including C_G, C_{GS}, C_{GD}, C_{SB}, and C_{DB} (proportional to W)

An implied dependence through the device transconductance is

- The maximum current through the transistor

since increasing the channel width creates a larger channel cross-section.

MOSFETs form the basis of digital CMOS circuits because they provide the controlled switching functions. However, the electrical characteristics and parasitics constitute the limiting factors on the speed that a logic circuit can operate.

CMOS logic design is relatively straightforward. It uses the basics of Boolean switching logic as applied to nFETs and pFETs, and results in a methodology where the logic is determined solely by the placement of the transistors. Changing the size of any (or all) transistors in a basic CMOS logic gate does not affect the logic operation it performs. The speed of the switching, however, is determined entirely by the electrical parameters of the transistors. This is where layout becomes critical.

● ● ● ● ● ● ● ● ● ● ● ● ● ● ●

6.6   References

[6.1]   Baker, R. J., Li, H. W., and Boyce, D. E., *CMOS Circuit Design, Layout, and Simulation*, Piscataway, NJ: IEEE Press, 1998.

[6.2]   Pierret, R. F., *Semiconductor Device Fundamentals*. Reading, MA: Addison-Wesley, 1995.

[6.3]   Uyemura, J. P., *CMOS Logic Circuit Design*. Norwell, MA: Kluwer Academic Press, 1999.

[6.4]   Uyemura, J. P., *Introduction to VLSI Circuits and Systems*. New York: John Wiley & Sons, 2002.

[6.5]   Wolf, W. H., *Modern VLSI Design, Third Edition*. Upper Saddle River, NJ: Prentice-Hall, 2002.

6.7   Exercises

6.1   Consider an MOS capacitor with a gate-oxide thickness of 30 Å. Find the value of the oxide capacitance per unit area, C_{ox}, in units of fF/μm^2. Assume that $\varepsilon_{ins} = \varepsilon_{ox} = 3.9\,\varepsilon_o$ with $\varepsilon_o = 8.854 \times 10^{-14}$ F/cm being the permittivity of free space.

6.2   An nFET has a 3.5 nm gate oxide. The electron mobility is $\mu_n = 550$ cm^2/V-sec. The gate dimensions are $W = 0.80$ μm and $L = 0.18$ μm.

(a)   Find the gate capacitance, C_G, in fF.

(b)   Calculate the device transconductance, β_n, in units of μA/V^2.

6.3   A p-channel MOSFET has a gate oxide that is 40 Å thick with a hole mobility of $\mu_p = 200$ cm^2/V-sec. The channel length is 0.25 μm and $W = 1$ μm.

(a)   Find the gate capacitance, C_G, in fF.

(b)   Calculate the device transconductance, β_p, in units of μA/V^2.

6.4   Consider an nFET where $\mu_n = 550$ cm^2/V-sec, $t_{ox} = 30$ Å, $W = 0.60$ μm, and $L = 0.18$ μm. The threshold voltage is $V_{T0} = 0.40$ V and the power supply is 2.5 V. Use the square law models to answer the following questions.

(a)   Find the current with $V_{GS} = 1$ V, $V_{DS} = 1.25$ V, and $V_{SB} = 0$ V.

(b)   Calculate the FET resistance using Equation (6.37).

(c)   Calculate the FET resistance using Equation (6.40).

6.5   A pFET has $\mu_p = 190$ cm^2/V-sec, $t_{ox} = 30$ Å, $W = 0.60$ μm, and $L = 0.18$ μm. The threshold voltage is $V_{T0pn} = -0.45$ V and the power supply is 2.5 V. Use the square law models to answer the following questions.

(a)   Find the current with $V_{SG} = 1$ V, $V_{SD} = 1$ V, and assume zero body bias.

(b)   Calculate the FET resistance using Equation (6.43).

MOSFET Modeling with SPICE

Analytical models are limited in their ability to accurately predict the behavior of MOS-FETs under time-varying conditions. The problem lies in the complexity of submicron device physics. It is not possible to write down a simple set of equations that can be used in hand calculations. The obvious solution is to move to the level of computer simulations. In this chapter, we will examine some basic CAD models of FETs to learn more about the behavior of transistors.

● ● ● ● ● ● ● ● ● ● ● ● ● ● ·· ·

7.1 SPICE Levels

A circuit simulation based on SPICE is capable of yielding reasonably accurate simulations, depending upon the models used. In standard SPICE syntax, a MOSFET in Figure 7.1 is described by the text line

Mname ND NG NS NB Mod_Name <Parameters>

where **Mname** is the name or label given to the particular device. The node numbers/labels are specified by **ND**, **NG**, **NS**, and **NB** for the drain, gate, source, and bulk, respectively, and **Mod_Name** refers to a specific **MODEL** statement. The **<Parameters>** listing in the device declaration is optional, but usually consists of W and L, in addition to the area and perimeter of the drain and source regions (AD, PD, AS, PS). It is important to remember that SPICE uses strict MKS units, so all lengths are in meters.

The **MODEL** listing provides the general processing parameters. For an nFET, it has the general form

.MODEL Mod_Name NMOS LEVEL=X < Parameters>

Figure 7.1:
Basic SPICE
parameters

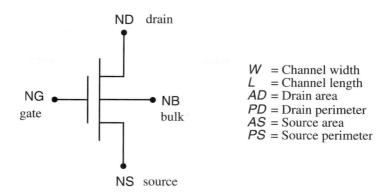

where **NMOS** implies an n-channel MOSFET. To characterize a pFET, the keyword changes to **PMOS**. The equations that are used to characterize the device are specified by the **LEVEL=X** specification. The **LEVEL** assignment also determines the **Parameters** that are needed for the simulation. Although there is some consistency among SPICE simulators for classical **Level 1–4** values, there are changes when the more advanced models are introduced.

7.1.1   Level 1 MOSFETs

The simplest MOSFET model in SPICE is given by **LEVEL = 1**. The Level 1 equations are basically square law equations. The saturation voltage is defined by $V_{\text{sat}} = V_{GS} - V_{Tn}$. For $V_{DS} \leq V_{\text{sat}}$, the transistor is in the triode region and is described by

$$I_D = \frac{\beta_n}{2}[2(V_{GS} - V_{Tn})V_{DS} - V_{DS}^2](1 + \lambda V_{DS}) \qquad (7.1)$$

The current in a saturated FET with a large drain–source voltage, $V_{DS} \geq V_{\text{sat}}$, is

$$I_D = \frac{\beta_n}{2}(V_{GS} - V_{Tn})^2(1 + \lambda V_{DS}) \qquad (7.2)$$

The factor $(1 + \lambda V_{DS})$ has been added to the triode expression, even though channel-length modulation occurs only in a saturated FET. This is done to provide a continuous function as the device changes regions.

Table 7.1 lists five of the primary input variables for a Level 1 simulation. These are easy to correlate to theory because they use obvious notation. Level 1 modeling accounts for all basic phenomena, such as calculating the threshold voltages from basic processing parameters, including nonlinear capacitor effects, the difference between L and L_{eff}, and body-bias effects. However, since the basic equations are for long-channel devices, they

TABLE 7.1 Level 1 SPICE MOSFET parameters

Parameter	Symbol	Units	Definition
VTO	V_{T0}	V	Threshold voltage
TOX	t_{ox}	m	Oxide thickness
UO	μ_o	m²/ V-sec	Low-field mobility
PHI	$2|\phi|$	V	Surface potential
GAMMA	γ	V$^{1/2}$	Body-bias coefficient

are only slightly better than the analytical square law models. It is useful for seeing overall behavior of circuits since the simulations run very quickly.

The next higher set of MOSFET equations are contained in Level 2 expressions. These are more complicated and take into account some of the details of the bulk charge distribution that are ignored in the Level 1 model and some small device effects. Extending the equation set leads to the Level 3 model, which further improves the accuracy of small device calculations without increasing the simulation time too much.

The most accurate SPICE model for submicron MOSFETs is **Berkeley Short-Channel IGFET Model (BSIM)**[1]. The first release (BSIM1) was developed in the mid-1980s for modeling FETs with a channel length of about 1 μm or larger. BSIM1 introduced many empirical curve-fitting parameters to help with scaling issues, and BSIM2 improved on the idea somewhat. The third release, BSIM3, was based on a 2-dimensional analysis of both geometrical and electromagnetic effects in submicron devices. **BSIM3.v3** evolved to use a single "master" equation to describe the current, with the input factors calculated from fundamental relationships. The latest release (BSIM4 at this writing) can give very accurate calculations for submicron devices, but it does require a somewhat large number of parameters.

Since our interest is directed towards the engineering aspects of CMOS layout and simulation, we will not delve deeper into the origin of the modeling equations. Our approach will be more pedestrian, one that concentrates on using SPICE as a CAD tool to understand the interplay between layout and performance. The only assumptions that we will make deal with the expected accuracy of the various models. In particular, for submicron CMOS circuits,

- Level 1 modeling is the simplest and can be correlated to hand calculations, but is the least accurate
- Level 2 and 3 models are more accurate than Level 1
- BSIM provides the most precision of the three model groups

This will allow us to understand some of the more intricate aspects of chip design.

---

1. IGFET stands for Insulated-Gate FET, another name for what we now generally call a MOSFET.

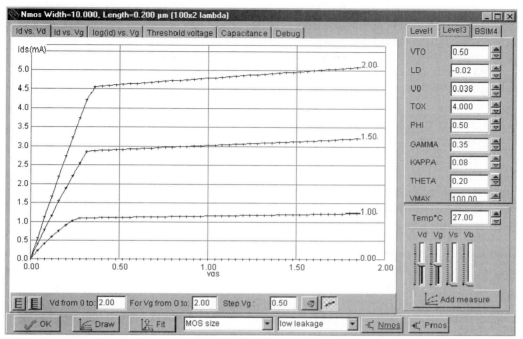

Figure 7.2:   Level 3 MOSFET characteristics

● ● ● ● ● ● ● ● ● ● ● ● ●
7.2   MOSFET Modeling in Microwind

Microwind has an integrated SPICE program that allows you to simulate the electrical circuits designed with the layout editor. There are three MOSFET models included with the simulator: Level 1, Level 3, and BSIM4. The modeling equations have been simplified somewhat by limiting the number of input parameters. This was done to incorporate data from a large number of foundry processes, while still providing good simulation results. The Microwind equations and parameter list for Level 3 and BSIM4 analyses are provided in the last section of this chapter. The user interface allows you to select the Level for the simulation, review the results, then change the Level if desired. In this section, we will use it to study the characteristics of MOSFETs from the CAD viewpoint.

The **DC MOSFET** curves can be viewed by clicking the on-screen button shown; the button is located on the right side of the **Menu** bar. This action displays a family of nFET curves for I_D versus V_{DS} using a Level 3 model as a default. The screen dump in Figure 7.2 shows the appearance of the screen using the *cmos018.rul* technology file. The parameter list on the right side of the screen gives the Level 3 values used to generate the curves. The device size is given at the top of the screen as $(W/L) = (10/0.2) = 50$. Each curve in the family is for the gate–source voltage indicated.

The screen has several features that allow you to vary the device being simulated. First, note that the values of the SPICE device-modeling parameters can be increased or decreased using the red (up) and green (down) arrows next to each value. The aspect ratio is shown at the default value, but it can be changed to another standard value using the **MOS size** adjustment at the bottom center of the screen. You can adjust the values of V_{DS}, V_{GS}, and the step size, ΔV_{GS}, by typing the values for the options

Vd from 0 to:

Vg from 0 to:

Step Vg:

in the indicated areas. The "slide bars" in the lower-right portion of the screen allow you to create a curve for any voltage combination. Sliding the **Vg** bar up or down adjusts the gate voltage; once the point is at the desired value, slide the **Vd** bar to generate the curve. The **Add measure** button gives a window where you can select data from several measured devices; the measured curves are then overlaid on top of the simulated plots for comparison. You can adjust the device type from **low leakage** to **high speed** or **high voltage.** The small yellow button enables the memory mode and allows you to save a screen and superpose another on top of the displayed curves. Finally, the **Pmos** button allows you to plot pFET curves that can be manipulated in the same manner.

One nice feature of the Microwind MOS curve screen is the ability to change the level of SPICE modeling by selecting the tab on top of the parameter list. For reference, note that the Level 3 plot shows a drain current of about 5.3 mA with V_{DS} and V_{GS} both around 2 V. Pushing the **Level1** button changes the modeling level and creates the new display shown in Figure 7.3. In addition to the shape of the curves being very different, the current predicted by the simplified Level 1 equation is about 21 mA for the same V_{DS} and V_{GS} values. This illustrates the error introduced by the simpler equations. Note that the SPICE parameter list has changed, corresponding to the difference between the two models.

The **BSIM4** curves are displayed by selecting the appropriate tab. This gives the plot shown in Figure 7.4. These are assumed to be the most accurate and predict a drain current of about 5.8 mA for V_{DS} and V_{GS} around the 2 V values. The comparison shows that the Level 1 model overestimates the current by over 350%, while the Level 3 values are much closer. This example illustrates why one must be careful when using the square law equations for quick estimates. Level 3 models are the default in Microwind, since they allow fast computation without excessive error.

The default MOSFET plot is **Id vs. Vd**, but other transistor characteristics can be accessed using the tabs at the top of the plot. The current as a function of gate voltage is obtained by clicking on the **Id vs. Vg** tab. This displays the plot shown in Figure 7.5. The family of curves is generated by varying the bulk voltage. The same information is contained in the plot of **log(Id) vs. Vg** of Figure 7.6, but the logarithmic current scale is more useful for estimating leakage currents that flow even when $V_{GS} < V_{Tn}$; this is called **subthreshold current**, since the gate voltage is below the threshold value. Note that the zero body-bias threshold voltage is specified as $V_{T0} = 0.50$ V. The plot of I_D as a function

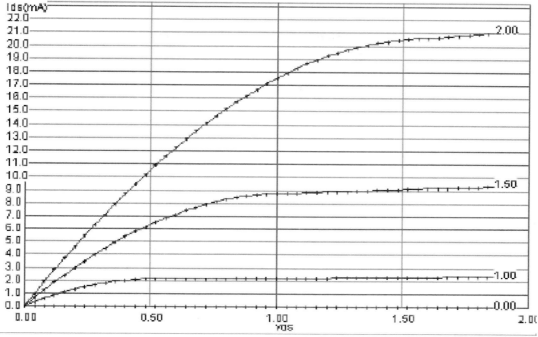

Figure 7.3: Level 1 SPICE modeling

of V_{GS} shows that the turn-on characteristics are not sharp, but continuous, as expected on physical grounds. For subthreshold conduction, we define the **subthreshold slope**, S, by

$$S = \left[\frac{d}{dV_{GS}} \ln(I_D) \right]^{-1} \tag{7.3}$$

such that the subthreshold current is approximated by

$$I_{\text{sub}} = I_x \left(\frac{W}{L} \right) [1 - e^{-(V_{DS}/V_{\text{th}})}] \, e^{-(V_{GS} - V_{Tn})/S} \tag{7.4}$$

with V_{th} the thermal voltage. Note that subthreshold current increases with the bulk voltage. Typical devices in modern technologies exhibit leakage-current levels on the order of a few nanoamperes per micron of width.

The last example we will look at here is the plot produced by the **Capacitance** tab. This brings up the plot of **gate capacitances** shown in Figure 7.7. The graph gives values of C_{GS} (the top curve) and C_{GD} (the bottom curve) as the drain–source voltage is

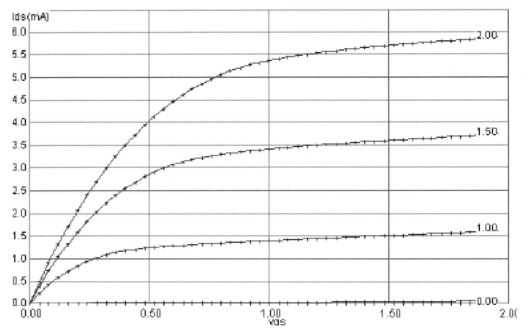

Figure 7.4: BSIM4 nFET characteristics

increased. This places the triode (non-saturation) region on the left, and the saturation region on the right side. The triode-region gate-related capacitances are computed from

$$C_{GS} = \frac{2C_G}{3}\left[1 - \left(\frac{V_{GS} - V_t - V_{\text{sat}}}{2(V_{GS} - V_t) - V_{\text{sat}}}\right)^2\right]$$

$$C_{GD} = \frac{2C_G}{3}\left[1 - \left(\frac{V_{GS} - V_t}{2(V_{GS} - V_t) - V_{\text{sat}}}\right)^2\right]$$

(7.5)

with V_t the **thermal voltage**.

Junction capacitances, C_{DB} and C_{SB}, are nonlinear and are computed from the standard expressions

$$C_{DB} = WX_{\text{drain}}\frac{C_j}{\left(1 - \frac{V}{\phi_o}\right)^{m_j}}$$

$$C_{SB} = WX_{\text{source}}\frac{C_j}{\left(1 - \frac{V}{\phi_o}\right)^{m_j}}$$

(7.6)

Figure 7.5:
Drain current as a
function of gate
voltage

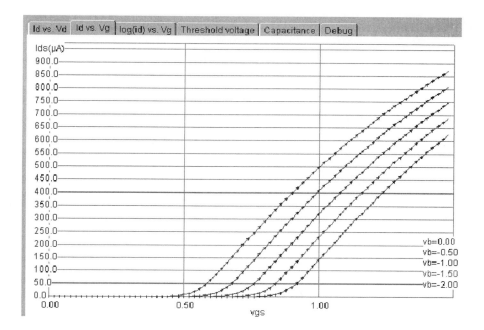

where X_{drain} and X_{source} are the extent in meters of the ndiff or pdiff regions beyond the gate, C_j is the zero-bias junction capacitance per unit area in (F/m^2), ϕ_o is the built-in voltage (**PB** or **potential barrier** in standard SPICE notation), and V is the **forward voltage** across the junction. In practice, $V < 0$, as the junctions are always kept in reverse bias.

Figure 7.6:
Subthreshold
current

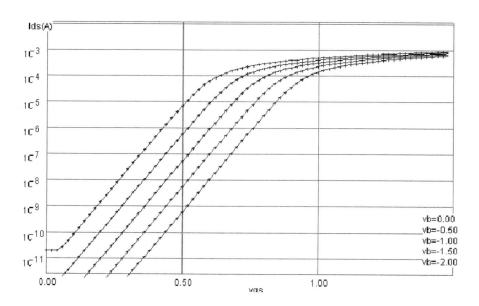

Figure 7.7:
Nonlinear gate
capacitances

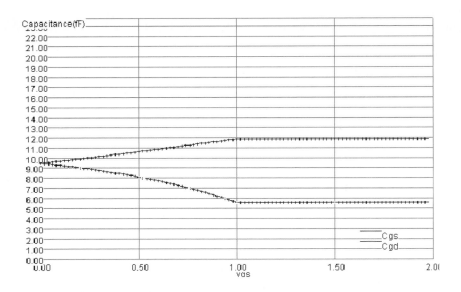

7.3 Circuit Extraction

A **circuit extractor** is a program that translates an integrated-circuit layout file into a netlist of electronic components. For example, mask overlays described by

(ndiff) ∩ (Poly)

are identified as nFETs such that

- **Poly** is the gate
- One side of **ndiff** is the drain
- The other side of **ndiff** is the source
- The p-type substrate is the bulk

An extractor will assign numbers (or labels) to each node, and determine **W** and **L** values for the geometry. The output for a single device would be a SPICE listing such as

M1 55 40 20 10 nFET W=1.0U L=0.25U

where **U** indicates microns. The extractor may also provide drain and source dimensions or parameters, (*AD*, *PD*, *AS*, *PS*). Circuit wiring is specified by tracing conducting lines (poly or metal) to every point and assigning the same node number to all points. Extracting an entire layout results in a SPICE listing for the electronic simulation.

Figure 7.8:
nFET layout
example

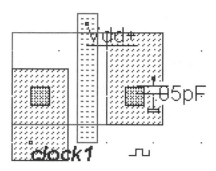

In this section, we will examine the characteristics of FETs when used as electrical switching elements using the extraction and simulation features of Microwind.

7.3.1   nFET Pass Characteristics

A pass nFET is one that is used to pass a signal voltage from one n^+ region to the other, with the gate acting to control the OPEN or CLOSED nature of the switch.

To begin this study, launch Microwind and then use the MOS generator to place an nFET on the work area. The example shown in Figure 7.8 was created in the *cmos018.rul* process with a channel width of $W = 1.0 \ \mu$m, but the L dimensions are not too important to the basic simulation. If you want to start from scratch, this is a good time to practice your FET drawing skills. Note that the drawing has features that are not automatically created in the layout. These are user-defined parameters that have been added to perform an electrical simulation.

General purpose **SPICE** buttons can be found at the top part of the **Palette** menu. If it is not visible in the work area, use the **Menu** button to activate it. A detailed view of the button arrangement is shown in Figure 7.9. The bottom row is dedicated to specialized SPICE functions and are identified with the following options:

Figure 7.9:
Palette Menu
buttons

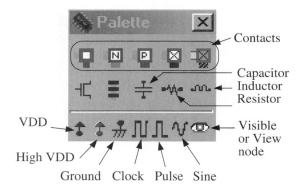

- V_{DD} provides a connection to the positive terminal of the power supply
- High V_{DD} (V_{DDH}) is for higher voltage FETs that may be found in the process
- Ground (or V_{SS}) connection
- Clock is an independent voltage source that provides a periodic square wave output that is referenced to ground
- Pulse is an independent voltage source that provides a pulse waveform and is referenced to ground
- Sine is an independent voltage source that provides a sine wave referenced to ground
- Visible or View Node makes the node visible in the SPICE simulation

To use a feature, click on a button and then point the mouse cursor to the desired node on the layout and click. In the case of a V_{DD} power supply or V_{SS} ground, a script and a node dot will appear to indicate that the connection has been made. Clicking on the other buttons will open a dialog window that allows you to see the details; closing the window activates the cursor, and you can connect the source to the circuit by clicking the node on the layout. For our example, we have added a Clock with default parameters to the left Metal/ndiff terminal, and a V_{DD} button to the gate poly to bias the nFET into conduction. A **virtual capacitor** with a value of 0.05 pF has been added to the right Metal1/diff. The device was added by first clicking the capacitor button on the **Palette Menu**, entering a value of 0.05 into the dialog window, and then clicking on the right node. Virtual elements are added to help the simulation, but are not physical devices. For future reference, note that virtual resistors and inductors can also be created using the indicated **Palette** window buttons. This completes the circuit.

Before starting the SPICE simulation, we must specify the nodes that we want to see in the output calculations. Nodes with DC voltages (V_{DD}, V_{DDH}, or ground/V_{SS}) normally are not shown. In the present case, the clock input and the voltage across the capacitor are important to show, since they illustrate the voltage-passing characteristics of the nFET. The node characteristics are displayed using the **View Node** button shown to the left. This actives the **Navigator** function. Pointing to any node in the layout and clicking on it opens up a window that has the general features portrayed in Figure 7.9. The electronics specialist will find this to be one of the most useful tools for tracing layouts. The upper part of the **Navigator** window gives the name (if specified), the total capacitance and resistance values, and the length. The **Node Properties** window includes all contributions at every layer that touches it, and the relevant parts of the layout will be highlighted, while the other sections remain transparent. Although the **Node Navigator** has many uses, we are most interested in using it to "activate" a node for viewing in SPICE. In Figure 7.10, the node will be visible, as indicated by the

View Node button

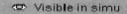

line. If the node is not included in the output list, then you will see

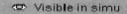

Figure 7.10:
Navigator Window

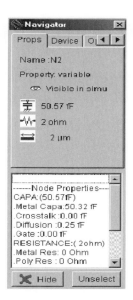

Just click on the bar and it will change to the Visible indicator. Check both the left (Clock) and right (Capacitor) sides of the FET and make both visible.

To start the SPICE simulation of the pass FET characteristics, use the **Run Simulate** button shown here, execute the menu command

Simulate ⇒
Run Simulation

Run Simulation
button

or use the keyboard sequence **[Ctrl][S]**. This action extracts the circuit to a SPICE file, opens a simulation window with the selected nodes displayed, and starts the simulation. For the case of the nFET pass transistor, the screen appears as in Figure 7.11; note that the screen indicates a Level 3 simulation model. The independent clock input has a voltage range from 0 to 2 V and a period of 1 μs. Since the nFET is biased ON with V_{DD} at the gate, the output voltage in the lower plot has the same period. However, with this level of modeling, the output voltage only ranges from 0 to 1.145 V, as obtained using the cursor measurement function. Although increasing the period will increase the capacitor voltage slightly (since the capacitor will have sufficient time to reach a fully charged state), the failure to attain the 2 V amplitude at the output is an intrinsic limiting characteristic of the nFET.

The threshold drop through an nFET can be understood using the illustration in Figure 7.12, where a high (logic 1) voltage of V_{DD} has been applied to the left input. A gate voltage of $V_G = V_{DD}$ turns the nFET ON and gives a good conducting channel to the right side. Summing voltages from the gate terminal to the right side gives

$$V_G = V_{GS} + V_{\text{out}} \qquad (7.7)$$

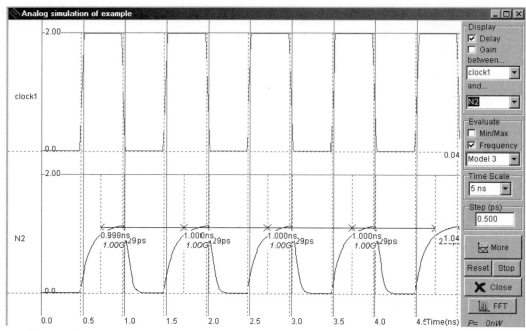

Figure 7.11: SPICE simulation of the pass transistor

Note, however, that the minimum gate–source voltage is $\min(V_{GS}) = V_{Tn}$, which is required to maintain the channel. Substituting gives

$$V_{DD} = V_{Tn} + V_{\text{out}} \tag{7.8}$$

or

$$
\begin{aligned}
V_{\text{out}} &= V_{DD} - V_{Tn} \\
&= V_{\text{max}}
\end{aligned}
\tag{7.9}
$$

as the maximum transmitted voltage. This says that an nFET can pass a logic 1 high voltage, but that the received signal is "weaker" by an amount V_{Tn}. This is called a **threshold-**

Figure 7.12:
Threshold drop

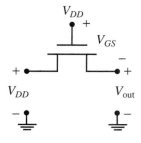

Figure 7.13:
Summary of nFET
pass characteristics

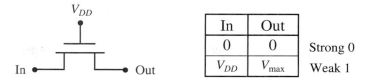

In	Out	
0	0	Strong 0
V_{DD}	V_{max}	Weak 1

voltage loss problem. When we include the source–body voltage V_{SB}, we see that body-bias increases the threshold voltage to a value larger than V_{Tn}, which reduces V_{max} even more.

The voltage pass characteristics of an nFET are summarized in Figure 7.13, where we have introduced the idea of "strong" and "weak" values. A "strong" value means that the signal can achieve the ideal logic voltage, while a "weak" transmission implies a change. Since switch models do not include the threshold-voltage loss, they must be used with care. A weak logic 1 indicates a lower electrical drive-capability, and this usually results in longer switching times.

7.3.2   pFET Pass Characteristics

A pFET is the opposite of an nFET. To examine the transmission characteristics, let us start with a new file in Microwind and use the MOS generator to create the pFET shown in Figure 7.14. The width only affects the switching times, and is not critical to this example. Note that we have added a clock to the left side, and have biased the gate at V_{SS} (= ground) to ensure that the pFET is conducting. A 0.05-pF capacitor has been added to the right side as a load. Before running the simulation, use the **Navigator** to make the left and right nodes visible in the simulation.

If you used the MOS generator to create the pFET, then activating the simulation results in the **Warning** window in Figure 7.15. This is because the n-well needs to be

Figure 7.14:
pFET layout with
clock and bias

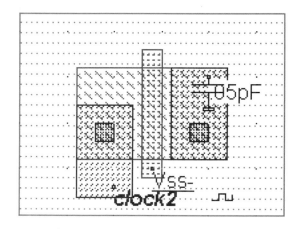

Figure 7.15:
Well bias
(polarization)
warning

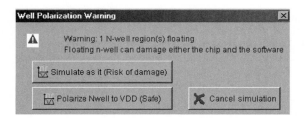

biased ("polarized" in the jargon of Microwind) to V_{DD} to ensure that the pn junctions are reverse biased. You can solve this by adding an explicit V_{DD} connection to the drawing, or just push the **Polarize Nwell to VDD** button to allow the simulation to run. Using the button approach then completes the extraction process and gives the simulation results shown in Figure 7.16. The clock still ranges from 0 to 2 V as in the nFET case, but now the output voltage is limited to the range of about 1.265 to 2 V (aside from the startup value). In other words, the pFET cannot pass an ideal logic 0 voltage of 0 V, which is exactly the opposite of the nFET.

In the case of a pFET, the threshold "loss" is the inability to pass a strong logic 0 voltage of 0 V. To understand this, consider the transistor shown in Figure 7.17, where the gate has been grounded so that $V_G = 0$. Summing the voltages gives

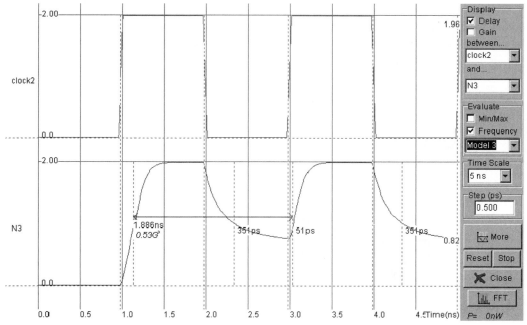

Figure 7.16: pFET pass characteristics

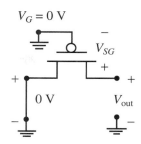

Figure 7.17:
pFET logic 0

$$V_G + V_{SG} = V_{\text{out}} \tag{7.10}$$

Since the minimum source–gate voltage is $|V_{Tp}|$, we have

$$0 + \min(V_{SG}) = V_{\text{out}} \tag{7.11}$$

so that the minimum output voltage at the output is

$$V_{\text{min}} = |V_{Tp}| \tag{7.12}$$

which is noticeably larger than the ideal logic 0 value of 0 V. Body bias increases V_{min} even more. Thus, the pFET can pass a strong logic 1 (V_{DD}), but only a weak logic 0, voltage.

The overall pFET pass characteristics are summarized in Figure 7.18. We see that they are exactly the opposite of the nFET: a pFET can pass a strong logic 1, but only a weak logic 0. In other words, nFETs and pFETs complement each other. This is the electrical basis for complementary design in CMOS. When voltage levels are important, we design the circuits so that

- nFETs are used to pass only 0 V levels
- pFETs are used to pass only V_{DD} levels

This philosophy is used to construct the basic family of CMOS logic gates. In more advanced digital-design techniques, the threshold-voltage effects can be dealt with in an acceptable manner by modifying the next stage in the logic cascade.

Figure 7.18:
Summary of pFET
pass characteristics

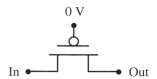

In	Out	
0	V_{min}	Weak 0
V_{DD}	V_{DD}	Strong 1

Figure 7.19:
SPICE extraction
dialog window

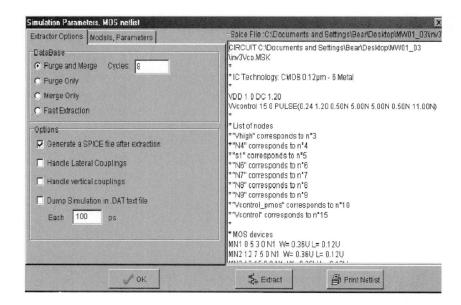

7.3.3 Using Other SPICE Simulators

Extracted netlists produced by Microwind can be exported to a different (user supplied) SPICE simulator if desired.

You must start with an open layout drawing, then use the Menu command sequence

> **File** ⇒
> > **Convert Into...**⇒
> > > **SPICE netlist**

This will open up a window entitled **Simulation Parameters**, as illustrated in Figure 7.19 The SPICE listing itself is contained in a sub-window on the right side. You can scroll down the list using the down arrow. Selecting the option

Generate a SPICE file after extraction

and pushing the on-screen **Extract** button will create a text file in the Microwind folder with the same name and a .cir extension. This may be copied into the text input of a different SPICE program. It is important to note that you may have to do some syntax editing and transfer values to conform to the input requirements.

7.4   Microwind Level 3 and BSIM4 Equations

Microwind uses simplified equations to describe MOSFETs in SPICE. The Level 1 expressions have already been presented. The Level 3 and BSIM4 equations are summarized here along with the relevant parameter lists for your reference.

7.4.1   Level 3

The Microwind current equation for $V_{GS} \geq V_{on}$ is given by

$$I_D = K_{\text{eff}} \frac{W}{L_{\text{eff}}} (1 + \kappa V_{DS}) V_{DE} \left[(V_{GS} - V_T) - \frac{1}{2} V_{DE} \right] \tag{7.13}$$

where

$$V_{on} = 1.2 V_T \tag{7.14}$$

is the ON voltage and

$$V_T = V_{T0} + \gamma (\sqrt{2|\phi| - V_B} - \sqrt{2|\phi|}) \tag{7.15}$$

is the threshold voltage. Other factors are

$$V_{DE} = \min(V_{DS}, V_{D\text{sat}}) \tag{7.16}$$

$$V_{D\text{sat}} = V_c + V_{\text{sat}} - \sqrt{V_c^2 + V_{\text{sat}}^2} \tag{7.17}$$

$$V_{\text{sat}} = V_{GS} - V_T \tag{7.18}$$

$$V_c = \frac{L_{\text{eff}}}{0.06} v_{\text{max}} \tag{7.19}$$

where the effective channel length is

$$L_{\text{eff}} = L - 2L_D \tag{7.20}$$

TABLE 7.2 Level 3 Parameters

Parameter	Symbol	Units	Definition
VTO	V_{T0}	V	Long-channel threshold voltage
TOX	t_{ox}	m	Oxide thickness
UO	μ_O	m^2/V-sec	Low-field mobility
PHI	$2|\phi|$	V	Surface potential
GAMMA	γ	$V^{1/2}$	Body-bias coefficient
KP	k'	A/V^2	Process transconductance
PHI	$2|\phi|$	V	Surface potential
KAPPA	κ	V^{-1}	Saturation field factor
LD	LD	m	Lateral diffusion length
VMAX	v_{max}	m/sec	Maximum drift velocity
NSS	$1/nkT$	V^{-1}	Subthreshold factor
THETA	θ	V^{-1}	Mobility degradation factor

and

$$K_{\text{eff}} = \frac{k'}{[1 + \theta(V_{GS} - V_T)]} \tag{7.21}$$

Subthreshold current is expressed by

$$I_D = I_D(V_{\text{on}}, V_{DS})e^{(V_{GS} - V_{\text{on}})/V_{\text{th}}} \tag{7.22}$$

where V_{th} is the thermal voltage. The parameter listing is given in Table 7.2.

7.4.2 BSIM4 Equations

Microwind uses a simplified BSIM4 equation set to allow fast simulations. The full BSIM4 description uses around 200 parameters; the Microwind implementation is based on about 20 of the most important values. The basic equations are presented here.

The threshold voltage in the BSIM model is given by

$$V_T = V_{T0} + K_1(\sqrt{2|\phi| - V_{BS}} - \sqrt{2|\phi|}) - K_2 V_{BS} + \Delta V_{T,\,\mathrm{SCE}}$$
$$+ \Delta V_{T,\,\mathrm{NULD}} + \Delta V_{T,\,\mathrm{DIBL}} \qquad (7.23)$$

where $\Delta V_{T,\mathrm{SCE}}$ is the **reduction due to short channel effects**, $\Delta V_{T,\mathrm{NULD}}$ is due to nonuniform lateral doping effects, and $\Delta V_{T,\mathrm{DIBL}}$ is the change in the threshold voltage arising from **Drain-Induced Barrier Lowering (DIBL)**.

The effective mobility is

$$\mu_{\mathrm{eff}} = \frac{\mu_o}{1 + (u_A + u_C V_{BS,\,\mathrm{eff}})\left[\dfrac{1}{t_{\mathrm{ox}}} V_{GS,\,\mathrm{eff}} + 2(V_{T0} - V_{FB} - \phi_S)\right]^{EU}}, \qquad (7.24)$$

where $EU = 5/3$ for nFETs, and $EU = 1$ for pFETs. The current flow is computed from

$$I_{D0} = \mu_{\mathrm{eff}} \frac{\varepsilon_{\mathrm{ox}}}{t_{\mathrm{ox}}} \frac{W_{\mathrm{eff}}}{L_{\mathrm{eff}}} V_{GS,\,\mathrm{eff}}\left[1 - \frac{A_{\mathrm{bulk}} V_{DS,\,\mathrm{eff}}}{2 V_{GS,\,\mathrm{eff}} + 4 V_t}\right]\left[\frac{V_{DS,\,\mathrm{eff}}}{1 + (V_{DS,\,\mathrm{eff}}/(v_{\mathrm{sat}} L_{\mathrm{eff}}))}\right], \qquad (7.25)$$

In this expression,

$$L_{\mathrm{eff}} = L - 2L_{\mathrm{int}} \quad \text{and} \quad W_{\mathrm{eff}} = W - 2W_{\mathrm{int}} \qquad (7.26)$$

are the effective channel length and width, respectively. The simplified model employs L_{int} and W_{int} as the correction factors that are user-defined. The thermal voltage is denoted as $V_t = (kT/q)$, and the effective device voltages, $V_{GS,\mathrm{eff}}$ and $V_{DS,\mathrm{eff}}$, are used to provide a smooth transition as the mode of operation changes. The Microwind BSIM4 model also provides for temperature dependences and capacitances in the standard manner.

The main parameters are listed in Table 7.3. A detailed discussion of the BSIM4 equation set and parameters is beyond the scope of this book.

7.4.3   Microwind Parameter Entry

The SPICE parameters are contained in the .rul file that defines the technology information. The *cmos018.rul* file is listed in Appendix B as an example. Examining the listing shows that parameters are given for each SPICE modeling level. In the listing, a prefix of "l3" indicates a Level 3 parameter, while "b4" indicates a BSIM3 value.

TABLE 7.3 Microwind BSIM4 parameters

Parameter	Symbol	Units	Definition
VTO	V_{T0}	V	Long-channel threshold voltage
VFB	V_{FB}	V	Flatband voltage
TOX	t_{ox}	m	Oxide thickness
K2	K_2	$V^{1/2}$	2nd-order body-bias factor
DVT0	DV_{T0}	V	1st-order short-channel factor
DVT1	DV_{T1}	V	2nd-order short-channel factor
LPE0	LPE_0		Lateral non-uniform doping parameter
ETA0	E_{TA0}		DIBL coefficient
NFAC	N		Subthreshold turn-on factor
UO	μ_O	m^2/V-sec	Low-field mobility
UA	μ_A	m/V	Vertical-field mobility factor
UC	u_C	V^{-1}	Body-bias mobility factor
PSCBE1	P_{SCBE1}	V/m	1st substrate-induced body-bias factor
PSCBE2	P_{SCBE2}	V/m	2nd substrate-induced body-bias factor
VSAT	v_{sat}	m/sec	Saturation velocity
WINT	W_{int}	m	Channel-width offset parameter
LINT	L_{int}	m	Channel-length offset parameter
KT1	K_{T1}	V	Temperature coefficient (V_T)
UTE	u_{TE}		Temperature coefficient (μ_O)
VOFF	V_{off}	V	Subthreshold offset voltage
PCLM	P_{CLM}		Channel-length modulation parameter

The syntax is best understood by example. The Level 3 nMOS parameters are shown below where the asterisk is parsed as a comment.

```
*
* Nmos Model 3 parameters
*
NMOS
l3vto = 0.5
l3v2to = 0.34
l3v3to = 0.7
l3vmax = 100e3
l3gamma = 0.35
l3theta = 0.2
l3kappa = 0.08
l3phi = 0.5
l3ld = -0.02
l3u0 = 0.038
l3tox = 4e-9
l3nss = 0.04
```

These can also be seen in the **MOS Characteristics Simulation** window that plots the transistor I–V curves.

Parasitic device capacitances are included in various parts of the file with the main contributions shown here:

```
*
* Parasitic capacitances
*
cpoOxyde= 4600 (Surface capacitance
*Poly/Thin oxide aF/µm2)
cpobody = 80   (Poly/Body)
...
cgsn = 500 (Gate/source capa of nMOS)
cgsp = 500
*
```

The .rul file also contains information on other parasitics for the circuit simulations.

The values can be modified as needed, but it is recommended that a new .rul file be created for each process to be characterized. It is important to note that Microwind derives all of its parameters from the selected .rul file through the **Select Foundry** command, so that a customized description should include all important parameters, such as design rules and parasitic line capacitances.

7.5 References

[7.1] Cheng, Y., and Hu, C., *MOSFET Modeling and BSIM3 User's Guide*. Norwell, MA: Kluwer Academic Press, 1999.

[7.2] Fjeldly, T. A., Ytterdal, T., and Shur, M., *Introduction to Device Modeling and Circuit Simulation*. New York: John Wiley & Sons, 1998.

[7.3] Foty, D., *MOSFET Modeling with SPICE*. Upper Saddle River, NJ: Prentice Hall, 1997.

[7.4] Sicard, E. *Microwind & DSCH User's Manual*. Toulouse, France: 2002.

7.6 Exercises

7.1 What is the maximum voltage that can be passed through the two-series nFETs shown below when $V_{in} = V_{DD}$? What is the minimum voltage when $V_{in} = 0$ V?

Problem 7.1

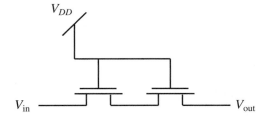

7.2 What is the minimum voltage that can be passed through the two-series pFETs shown below when $V_{in} = 0$ V? What is the maximum voltage when $V_{in} = V_{DD}$?

Problem 7.2

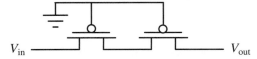

7.3 Compare the device curves for low-leakage and high-speed devices. What are the differences?

CMOS Logic Gates—Design and Layout

CMOS is the primary technology used for high-density VLSI logic chips, such as microprocessors and digital signal processors. CMOS logic gates are designed following a set of rules that associate logical operations with FET switching characteristics. The technique is straightforward to learn and provides a basis for building large logic networks.

This chapter introduces the topology and layout of circuits for basic logic operations. Emphasis is on the placement of FETs in both circuit diagrams and on silicon.

8.1 The Inverter

The inverter, or NOT gate, forms the basis for more complex CMOS logic circuits. The logic symbol and circuit schematic are shown in Figure 8.1. It is seen that the gate consists of a complementary nFET/pFET pair, Mn and Mp, that have been wired to ground, V_{SS}, and V_{DD}, respectively. The input a is applied to the common gate connection, and the output $NOT(a) = \bar{a}$ is taken at the common node between the two transistors. The NOT operation "flips" the input bit, i.e.,

$$NOT(0) = 1$$
$$NOT(1) = 0$$

(8.1)

as summarized in the truth table. The CMOS circuit implements this function by using the FETs as switches that are controlled by the input a. The pFET, Mp, is wired between the output and the power supply, V_{DD}, while the nFET, Mn, is placed between the output and

Figure 8.1:
NOT gate

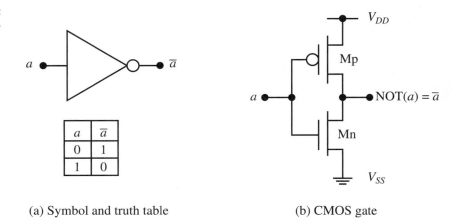

(a) Symbol and truth table

(b) CMOS gate

ground. Since the input a drives both gate terminals, we obtain complementary action such that

$a = 0$: the pFET is ON, while the nFET is OFF

$a = 1$: the pFET is OFF, while the nFET is ON

In switch terminology, an ON transistor acts like a closed switch, while an OFF transistor is an open circuit. This gives the operation portrayed in Figure 8.2. In Figure 8.2(a), the input $a = 0$ turns Mp ON so that the power supply voltage, V_{DD}, is transmitted to the output, while the nFET acts like an open circuit. This gives an output of Out = 1. Conversely, if $a = 1$ as in Figure 8.2(b), then the nFET acts like a closed switch that connects the output node to ground while Mp is OFF, yielding Out = 0.

It should be noted that the NOT operation is a result of the circuit topology, i.e., how the transistors are wired together. The pFET has been placed to conduct the logic 1, V_{DD},

Figure 8.2:
Inverter operation

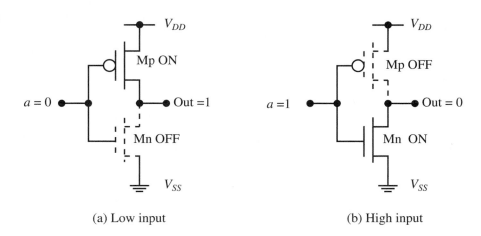

(a) Low input

(b) High input

Figure 8.3:
Voltage transfer
curve

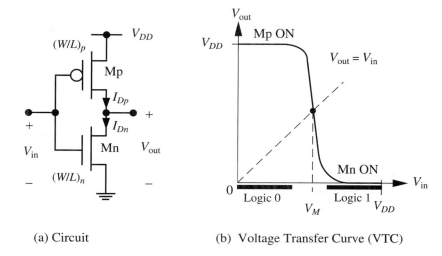

(a) Circuit   (b) Voltage Transfer Curve (VTC)

voltage to the output since it can pass a strong 1. Similarly, the nFET is used to conduct the ground voltage, $V_{SS} = 0$ V, to the output, since it can pass a strong 0.

8.1.1   Electrical Characteristics

The aspect ratios, $(W/L)_p$ and $(W/L)_n$, are the design variables for the circuit. Both DC and transient characteristics are set by these values.

The Voltage Transfer Curve

The DC properties of an inverter usually are shown using the **Voltage Transfer Curve** (VTC). This is a plot of the output voltage V_{out} as a function of the input voltage, V_{in}, from 0 to V_{DD}. Figure 8.3(a) shows the circuit, and the resulting VTC is drawn in Figure 8.3(b). VTC shows the logic 0 and logic 1 input voltage ranges for the circuit. This means, for example, that any voltage in the logic 0 range results in a high (logic 1) output.

The **output high** and **output low** voltages are given, respectively, by

$$V_{OL} = 0 \text{ V}$$
$$V_{OH} = V_{DD}$$

(8.2)

and the output **logic swing** is

$$V_L = V_{OH} - V_{OL}$$
$$= V_{DD}$$

(8.3)

This is called a **full-rail** output. When $V_{out} = V_{DD}$, the pFET is ON and the nFET is OFF. Conversely, when $V_{out} = 0$, the nFET is ON and the pFET is OFF. The full-rail output is thus a consequence of the transistor placement.

The voltages between the logic 0 and logic 1 input ranges do not represent any Boolean value and are classified as being undefined in the binary world. The transition from a high- to low-output voltage occurs because both FETs are conducting current. A good reference point is the midpoint voltage, V_M, that is in the transition region. As seen in the drawing, the midpoint voltage is where the VTC intersects the line $V_{out} = V_{in}$. It can be shown that both transistors are saturated at this point. We can estimate V_M using the **α-power current expressions**, with $I_{Dp} = I_{Dn}$, to write

$$\frac{\beta_p}{2}(V_{DD} - V_M - |V_{Tp}|)^\alpha = \frac{\beta_n}{2}(V_M - V_{Tn})^\alpha \tag{8.4}$$

where we have used $V_{SGp} = V_{DD} - V_{in}$ and $V_{GSn} = V_{in}$ and taken $\lambda = 0$ for simplicity. Taking the α-th root of both sides and rearranging gives

$$V_M = \frac{V_{DD} - |V_{Tp}| + \left(\frac{\beta_n}{\beta_p}\right)^{1/\alpha} V_{Tn}}{1 + \left(\frac{\beta_n}{\beta_p}\right)^{1/\alpha}} \tag{8.5}$$

with

$$\frac{\beta_n}{\beta_p} = \frac{k_n' \, (W/L)_n}{k_p' \, (W/L)_p} \tag{8.6}$$

This shows that V_M is determined by the ratio of device-transconductance values, which are in turn set by the aspect ratios. In other words, we can adjust the location of the midpoint voltage, V_M, using the ratio

$$\frac{(W/L)_n}{(W/L)_p} \tag{8.7}$$

in the layout. This allows us to control the logic 0 and logic 1 input voltage ranges.

Transient Characteristics

Every node in a CMOS circuit is capacitive. To model the switching characteristics of the inverter, we write the total-output capacitance

$$C_{out} = C_{FET} + C_L \tag{8.8}$$

as in the circuit shown in Figure 8.4. In this equation, C_{FET} represents the capacitance due to the inverter FETs, while C_L is the external load. The switching speed of the circuit is determined by the time interval needed to charge and discharge C_{out} as the input voltage, V_{in}, is changed.

Figure 8.4:
Output capacitance

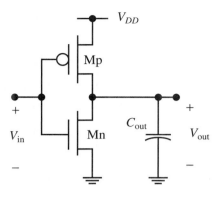

The simplest approach to modeling the transients is to assume that the input voltage is a simple square wave with a period, T. When $V_{in} = 0$, the output charges C_{out} to V_{DD}, while $V_{in} = V_{DD}$ discharges the capacitor to 0 V. This gives us the output rise time, t_r, and fall time, t_f, that are shown in Figure 8.5. Small values of t_r and t_f imply a fast circuit. Noting that the frequency of the input waveform is $f = 1/T$, we can define the maximum signal frequency by

$$f_{max} = \frac{1}{t_r + t_f} \tag{8.9}$$

This can be viewed as the upper limit for proper operation where the output is allowed sufficient time to stabilize to logic 0 and 1 voltages.

Figure 8.5:
Switching times

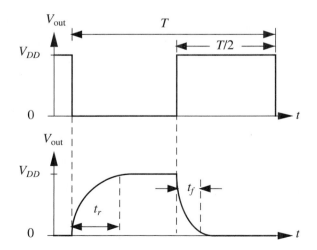

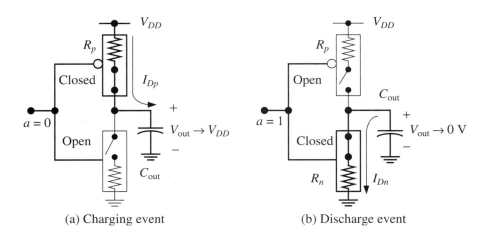

(a) Charging event (b) Discharge event

Simple expressions for the rise and fall times can be obtained by modeling the FETs as simple, controlled switches. In Figure 8.6(a), the input $a = 0$ closes the pFET switch while the nFET is open. This allows the current, I_{Dp}, to flow from the power supply and charge C_{out} to V_{DD} and is characterized by the time constant

$$\tau_p = R_p C_{out} \tag{8.10}$$

The rise time is usually written in the form

$$t_r = s_p \tau_p \tag{8.11}$$

where s_p is a **scaling factor** that accounts for the shape of the curve. For example, if we approximate the rising voltage as an exponential, then $s_p \approx 2.2$.

The fall time is estimated in the same manner. Figure 8.6(b) shows the case where $a = 1$, which closes the nFET switch and allows the capacitor to discharge through the nFET resistance, R_n. The time constant is now the product

$$\tau_n = R_n C_{out} \tag{8.12}$$

and the fall time is

$$t_f = s_n \tau_n \tag{8.13}$$

Note that since $C_{out} = C_{FET} + C_L$, both the rise and fall time can be expressed in terms of the external load capacitance by expressions of the form

$$t = t_0 + A C_L \tag{8.14}$$

Figure 8.7:
Switching times as
functions of
external load
capacitance

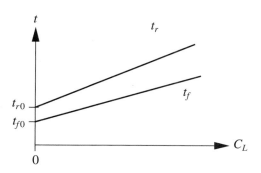

where t_0 is the delay with $C_L = 0$, and A is the slope of the time versus C_L. Figure 8.7 shows an example of the switching times. This illustrates an important point: Increasing the load slows down the gate.

The design aspect can be seen by recalling that the resistances are inversely proportional to the aspect ratios. The simple forms

$$R_n = \frac{2}{k'_n \, (W/L_n)(V_{DD} - V_{Tn})} \qquad , \qquad R_p = \frac{2}{k'_p \, (W/L_p)(V_{DD} - |V_{Tp}|)} \qquad (8.15)$$

show that using large W/L values decreases the FET resistance, which gives shorter time constants. There is a trade-off, however, since using large transistors increases the area. Also, Figure 8.4 shows that the inverter input looks capacitive, since we see two FET gate capacitances

$$\begin{aligned} C_{\text{in}} &= C_{Gn} + C_{Gp} \\ &= C_{\text{ox}}(WL)_n + C_{\text{ox}}(WL)_p \end{aligned} \qquad (8.16)$$

The input capacitance to the gate increases when we use large FETs, which makes it more difficult for the preceding stage to drive it.

8.1.2   Layout Considerations

Inverter layout is straightforward, since the circuit has only two FETs in it. However, it is worthwhile to examine some general ideas about creating a family of logic gates that are compatible with each other.

The first consideration is providing power supply, V_{DD}, and ground, V_{SS}, connections. Both will be supplied locally by horizontal Metal1 lines that are wide enough to accommodate contacts and vias. A simple calculation gives the width as 6 λ. This gives the basic structure of a **cell** shown in Figure 8.8. The vertical edge-to-edge spacing S between the VDD and VSS lines determines the amount of area that is allowed for transistors. Alternately, one can specify the **pitch** P, which is the spacing between the centerlines. The

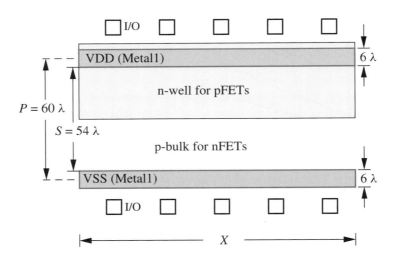

actual values vary with the design philosophy; we will select $P = 60\ \lambda$ corresponding to $S = 54\ \lambda$ to be consistent with pre-designed cells provided with Microwind for minimum line-width processes of 0.18 μm or larger. It is very important to keep the VDD and VSS widths and spacing the same for every cell that will be used to build a system. Note that Microwind technologies for 0.12 μm and smaller processes use $P = 50\ \lambda$ and $S = 44\ \lambda$ because of more compact spacing in the deep submicron design rules. If you are using the 0.12 μm technology file, adjust all VDD–VSS spacings accordingly.

An n-well is provided around the VDD rail to allow for pFETs and biasing. nFETs are placed by the ground (VSS) line. The extent X of the cell varies with the complexity of the logic circuit, with the inverter being relatively small in size. We have allowed for I/O (input/output) pads along the top and bottom to provide connection "channels" for interconnect wiring of the system components. Alternately, I/O points can be placed between the VDD–VSS lines, if room permits. Once dimensions for the n-well and the I/O strategy is selected, they should be used for every cell used to build a system.

Transistor Design and Placement

The aspect ratios of the transistors establish the electrical characteristics of the inverter. Selecting the actual size for each is a design choice that introduces the trade-offs.

The simplest layout is one where $(W/L)_n = (W/L)_p$. Using Equation (8.6) gives

$$\frac{\beta_n}{\beta_p} = \frac{k'_n}{k'_p} > 1 \tag{8.17}$$

Since the nFET resistance R_n will be smaller than the pFET resistance R_p, this choice gives $t_r > t_f$. Another approach is to set $\beta_n = \beta_p$ so that the aspect ratios are related by

$$(W/L)_p = \frac{k'_n}{k'_p}(W/L)_n \tag{8.18}$$

In this case, the pFET will be larger than the nFET. Typically, the process transconductance ratio (k'_n/k'_p) will be between three to six, and the problem reduces to selecting $(W/L)_n$. If $V_{Tn} = |V_{Tp}|$, then this design gives a **symmetric** inverter with $V_M = (1/2)V_{DD}$ and $t_r = t_f$. A first estimate of k'_n/k'_p can be obtained by taking the ratio of SPICE values $\mu_0(\text{nFET})/\mu_0(\text{pFET})$, even though this is not very accurate in submicron devices.

Let us step through the Microwind layout sequence for a symmetric inverter, assuming a value of $k'_n/k'_p = 2$ for simplicity. The aspect ratios are related by

$$(W/L)_p = 2(W/L)_n \tag{8.19}$$

and the channel lengths are the same minimum poly-width value of 2λ. If we choose $(W/L)_n = 4\lambda$, then $(W/L)_p = 8\lambda$. Our example will use the 0.18 μm technology described by the *cmos018.rul* file, where $2\lambda = 0.20~\mu$m. We will start by providing the VDD and VSS rails. These will be horizontal lines 6 λ high with edge-to-edge spacing of 54 λ. We will use $X = 30~\lambda$ as a first guess to the width and then adjust the size later.

> **Note:** *Remember to change this to an edge-to-edge spacing of 44 λ if you are using the 0.12 μm technology defined by the cmos012.rul specifications.*

The first step is to draw a Metal1 rectangle at the top of the screen with dimensions $(30~\lambda \times 6~\lambda)$; the dialog below the screen helps us obtain the desired sizing. This will be defined as VDD using the **Power Supply** button in the **Palette** window. Next, use the **Measure Distance** tool to create a vertical ruler below the VDD line to obtain the 54 λ spacing to the VSS line. The VSS line also has dimensions of $(30~\lambda \times 6~\lambda)$ and is identified as a ground connection. The Microwind layout with the temporary rulers is shown here.

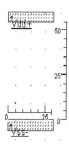

There are different techniques for orienting the transistors. We will redraw the circuit to that shown in Figure 8.9 to illustrate the method used here. This uses a common vertical polysilicon-gate line for both transistors. The output node must contact ndiff and pdiff regions, so it will be a Metal1. The actual output $\overline{a}$ will be parallel to the power-supply lines.

Figure 8.9:
Inverter FET
orientation

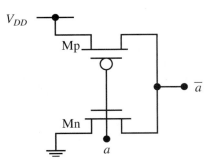

We will use the MOS Layout Generator for simplicity. First, create an nMOS transistor with $W = 4\,\lambda$, which is equivalent to $W = 0.8\ \mu$m in this technology. Remember that the MOS Layout Generator is activated using the FET button on the top portion of the **Pal-**

ette window. Point the cursor in the lower half of the drawing area near VSS and clicking gives an nFET there. The position can be moved using the editing tools. A pMOSFET is designed with $W = 8\,\lambda = 1.6\ \mu$m and placed near the VDD rail. It is convenient (but not necessary) to perform a vertical flip and then use the editing functions to align it with the VDD line.

The layout at this point should look something like that shown in Figure 8.10(a). Note that we have aligned the gates of the two transistors to allow for a single poly line. The final layout is shown in Figure 8.10(b). This has been obtained from the plot in Figure 8.10(a) by performing several minor edits on the drawing. These include

- Extending the poly-gate lines to meet and form a single vertical gate
- Extending the left-Metal1 FET lines to meet VDD (pFET) and VSS (nFET)
- Extending the right-Metal1 FET lines to create a common output with a horizontal extension
- Extending the top boundary of the n-well to get it under the VDD line
- Adding an n+ contact from VDD to the n-well to bias the pFET bulk
- Adding a p+ contact from VSS to the p-substrate to bias the nFET bulk

You should practice with the layout editor until your drawing looks similar to that shown. After working with the drawing and editing functions a short time, they will become quite

Figure 8.10:
NOT layout
sequence

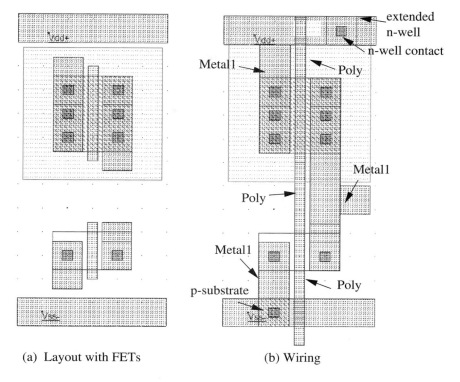

(a) Layout with FETs   (b) Wiring

familiar and easy to use. Connections can be checked using the Node Navigator. When you view a node, all of the polygons that are electrically connected will highlight and the characteristics will be listed in the **Navigator** window.

An electrical simulation can be obtained with the following steps:

- Place a clock on the poly-input line
- Add a virtual capacitor to the Metal1 output node. Use a value around 0.03–0.05 pF, and make the node visible using the Navigator
- Run the simulation. This will display the input clock and the inverter output

Figure 8.11 shows the waveforms obtained for this example. The load capacitor was chosen to be 0.05 pF, and the waveform was reduced in frequency from the default value to allow the output voltage to stabilize. The simulator can also be used to display the VTC by clicking on the lower tab labelled "Voltage vs. voltage." This gives the plot shown in Figure 8.12, which was generated using a Level 3 MOSFET model. This implies that there are some charge storage effects embedded in the results. These can be better seen by changing the model to BSIM4 using the controls.

This simple example illustrates the basic procedure needed to construct the layout for basic CMOS logic circuits. Once this becomes familiar, it easily can be extended to other logic circuits.

Figure 8.11:
Simulated inverter
transient response

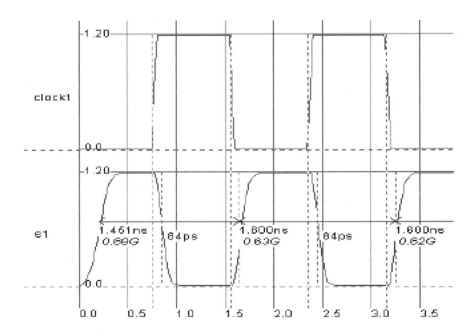

In this section, we have seen how to construct the basic CMOS inverter gate. This forms the basis for a class of circuits called the **static logic CMOS** design. A static gate is one where the output is well defined, so long as the inputs are specified and the power is applied. Static logic forms the basis for the great majority of digital CMOS chips. The primitive gates can be used to build larger functional units, which are in turn used as the basis for the design of large systems.

Figure 8.12:
The inverter VTC

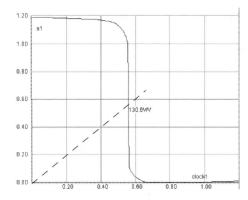

Figure 8.13:
Switch model of a
2-input CMOS
logic gate

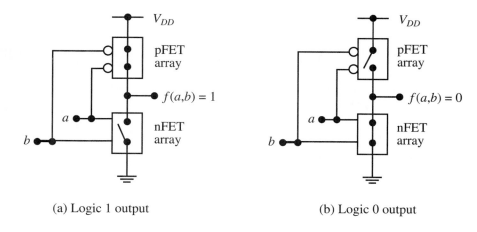

(a) Logic 1 output   (b) Logic 0 output

8.2   NAND and NOR Gates

Generic CMOS logic gates are based on the NOT circuit. The idea is quite simple. For an N-input gate, we use N complementary pairs that are arranged so that the output is either connected to V_{DD} through a pFET array, or connected to a ground through a group of nFETs. Figure 8.13 shows the idea for a 2-input gate. Each input a and b is connected to a complementary pair, with the nFETs and pFETs forming separate switch arrays. When the pFET switch array is closed, the nFET switch is open, giving an output of $f(a,b) = 1$, as shown in Figure 8.13(a). Conversely, when the nFET array is closed, the pFET group acts like an open circuit, and the output is $f(a,b) = 0$ [see Figure 8.13(b)]. The logical function $f(a,b)$ is determined entirely by the topology of the switching arrays, while the electrical characteristics result from the aspect ratios and associated parasitics.

8.2.1   Series and Parallel FETs

Consider a 2-input gate. The nFET array will consist of two nMOS transistors, and the pFET array will have two pMOS devices, such that each group functions as a switch from the top to the bottom. Given two transistors, there are two ways that we can connect them together: in series or in parallel. It is therefore worthwhile to investigate how to create the layout for these two possibilities.

Series-Connected FETs

By definition, two FETs in series have the same current. The circuit diagram in Figure 8.14(a) shows that the same current flows through both transistors. The layout is simplified because the point Y in the drawing can be translated into a common ndiff region that

Figure 8.14:
Series-connected
nFETs

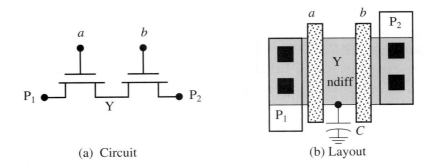

(a) Circuit (b) Layout

Figure 8.14:
Series-connected
nFETs

is shared by both FETs. This is shown in Figure 8.14(b). The two poly gates can be as close as 2λ together. The same idea is valid for pFETs that share a single pdiff region.

The real difficulty with using series transistors is that their resistances are added. This is illustrated in Figure 8.15. If both FETs have an aspect ratio, $(W/L)_n$, then each has a resistance, R_n, associated with it. When conduction is established from P_1 to P_2 with $a = b = 1$, then the total end-to-end resistance is

$$R_n + R_n = 2R_n \qquad (8.20)$$

If we string m nFETs in series, then the total resistance increases to mR_n. This can lead to a large time constant and slow down the switching. pFETs are even worse since, for equal aspect ratios, $R_p > R_n$. We therefore avoid long strings of series-connected transistors, but prefer nFETs over pFETs when we must use them.

Note that we can reduce the series resistance by increasing the channel width of each transistor. For example, if we double the aspect ratio of the transistors in Figure 8.14, then the series resistance is reduced to

$$\frac{1}{2}R_n + \frac{1}{2}R_n = R_n \qquad (8.21)$$

since each resistance is one-half of the original value. Larger transistors can therefore increase the switching speed.

Figure 8.15:
Series resistance
and capacitance

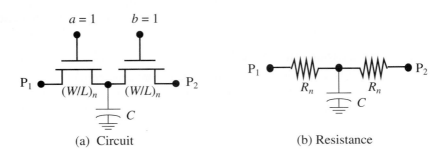

(a) Circuit (b) Resistance

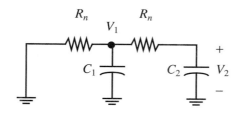

The existence of the capacitor, C, in between the two transistors causes many problems in high-speed design. Although this is a parasitic element, it will charge or discharge automatically when currents flow, which slows down the circuit. This is shown in Figure 8.16. If both capacitors initially are charged to a $t = 0$ voltage of $V_1(0) = V_X = V_2(0)$, then the discharge of C_2 can be described by the exponential function

$$V_2(t) \ = \ V_X \, e^{-t/\tau} \tag{8.22}$$

where the time constant is given by the sum of RC discharge path values

$$\tau \ = \ 2R_n C_2 + R_n C_1 \tag{8.23}$$

The first term is due to C_2 discharging through a total resistance of $2R_n$, while the second term is from C_1 discharging through one R_n. This is referred to as an **Elmore time constant**, and shows that series-connected FETs can be difficult to deal with. One approach is to use large aspect ratios to reduce the value of R_n.

Parallel-Connected FETs

When two FETs are wired in parallel, they have the same voltage across the drain–source terminals. Figure 8.17(a) illustrates a convenient way to view the parallel connection. We start with two transistors in series, but we connect the left and right ends (marked "B") together to form the common node, P_1. The other node, P_2, is taken from the common

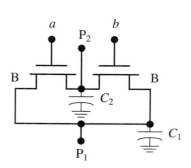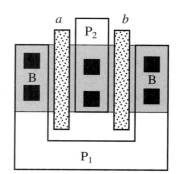

(a) Circuit   (b) Layout

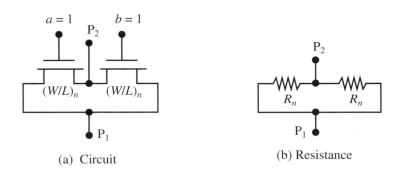

(a) Circuit   (b) Resistance

node. This is translated into a layout in Figure 8.17(b). In this case, the capacitances are important, since the value of C_1 consists of contributions from the large area ndiff regions seen on the left and right sides. The capacitance C_2 on node P_2 is smaller, since the ndiff area is smaller; this observation can be critical in high-speed design. Figure 8.18 shows how the resistances combine. Since the resistors are in parallel, the equivalent resistance between points P_1 and P_2 (when both FETs are ON) is

$$R_{P_1 - P_2} = \frac{1}{2} R_n \qquad (8.24)$$

so that the capacitance problem is of greater concern.

8.2.2   The NOR Gate

Consider a 2-input NOR (NOR2) gate. Denoting the inputs by a and b, the NOR operation

$$\text{NOR}(a, b) = \overline{a + b} \qquad (8.25)$$

is defined by

$$\text{NOR}(a, b) = 0 \text{ if either input is 1} \qquad (8.26)$$

so that NOR(a,b) = 1 iff $a = b = 0$. The standard logic symbols and truth table are shown in Figure 8.19; either a "+" or a "|" (pipe) will be used to denote the basic OR operation.

The CMOS gate is constructed by remembering that a 0 output indicates that the nFET switch is closed. The truth table shows that $\overline{a + b} = 0$ for the three combinations $(a,b) =$ (1,0), (0,1), and (1,1). For this to be true, the nFETs must be in parallel. Since the output is 1 for $(a,b) = (0,0)$, the pFETs are in series. This reasoning gives the NOR2 circuit shown in Figure 8.20, which has been drawn to clearly show that the nFETs Mna and Mnb are in parallel, while the pFETs Mpa and Mpb are in series. This style also makes the layout scheme obvious, since metal can cross over both poly and ndiff/pdiff regions without creating a contact.

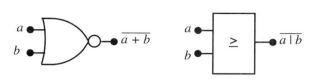

Figure 8.19:
NOR symbols and
truth table

a	b	$\overline{a+b}$
0	0	1
0	1	0
1	0	0
1	1	0

(a) Shape-specific   (b) IEEE   (c) Truth table

8.2.3 The NAND Gate

A NAND2 gate with inputs a and b implements the operation

$$\text{NAND}(a, b) = \overline{a \cdot b} \tag{8.27}$$

defined by

$$\text{NAND}(a, b) = 1 \text{ if either input is } 0 \tag{8.28}$$

Only the input combination $(a,b) = (1,1)$ gives NAND $(a,b) = 0$. The standard logic symbols and truth table are shown in Figure 8.21; note that the ampersand (&) is often used for the AND operation. Since both a and b must be 1 to produce a 0 output, the nFETs must be in series. Wiring the pFETs in parallel covers the remaining three cases, and yields the CMOS logic circuit in Figure 8.22. Note that the FET arrays are reversed from those used in the NOR2 gate. The nFET–pFET structures are examples of **series–parallel** arrangements in CMOS logic gates.

Figure 8.20:
NOR2 circuit
diagram

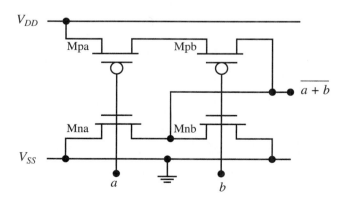

Figure 8.21:
NAND2 symbols
and truth table

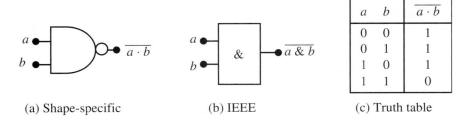

(a) Shape-specific   (b) IEEE   (c) Truth table

a	b	$\overline{a \cdot b}$
0	0	1
0	1	1
1	0	1
1	1	0

8.2.4   Layout

The NAND and NOR gates are logical duals of one another. This property translates to the series–parallel relationship between the nFETs and pFETs in the two gates. It is therefore instructive to examine the layout of both gates by starting with the same core of two nFETs and two pFETs. The distinction between the two gates is obtained by different wiring.

First, we again start with horizontal VDD and VSS (ground) lines that are 6 λ high. Since there are more FETs than we had for the inverter, the lines should be longer; with the device ratios selected in this example, horizontal lengths of about 40 λ will suffice.

The transistors can be added easily using the "multiple finger" option in the MOS Layout Generator. Select the macro by clicking on the **FET** symbol in the **Palette** window. Since we have two transistors, enter "2" in the dialog box for **Nbr of fingers** as shown in Figure 8.23. In layout terminology, a "finger" is a polysilicon-gate line. The Microwind macro can create a transistor with many fingers, and automatically provides ndiff or pdiff Contacts and Metal1 between every pair of gates. We will use a simple design where all transistors have a width of $W = 16\ \lambda = 1.6\ \mu$m in the 0.18 μm technology where $\lambda = 0.10\ \mu$m in the scalable rules. For placement purposes, you may find it easier to place the FET away from the VDD–VSS lines, then use the **Move** command to encircle the entire device and move it to the desired location. Figure 8.24 shows the layout when we create a 2-fingered nFET and move it close to VSS, and a 2-fingered pFET that is moved close to VDD. This core cell can be edited to make either a NOR2 or a NAND2 gate.

Figure 8.22:
NAND2 logic circuit

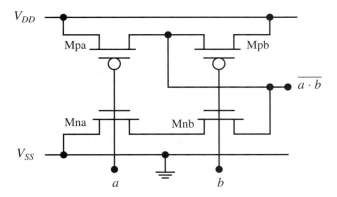

Figure 8.23:
Creating a 2-
fingered FET

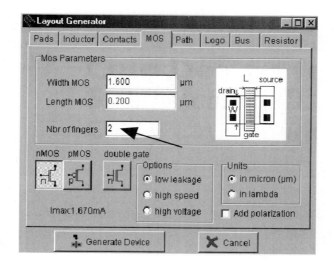

If you are following along on the PC, it would be a good time to save the layout using the Save As function; a good name might be "2_FETs." This will allow you to edit the layout and still have the core available for the other gate. The first edit that should be performed is to move the upper edge of the n-well so that it is underneath the VDD line. This will allow us to bias the n-well with the power supply. Some people may want to also perform a vertical flip operation on the pFET; this step reduces some of the editing, but it is not necessary. If you have chosen to construct the transistors manually, ensure that they are similar to the layout shown in the screen dump.

Figure 8.24:
Core cell for NOR2
and NAND2 logic
gates with Metal1
extensions to V_{SS}

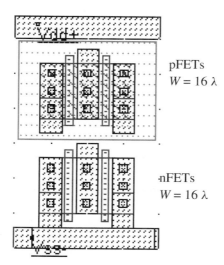

pFETs
$W = 16\,\lambda$

nFETs
$W = 16\,\lambda$

Figure 8.25:
NOR2 and NAND2
gates created from
the core cell

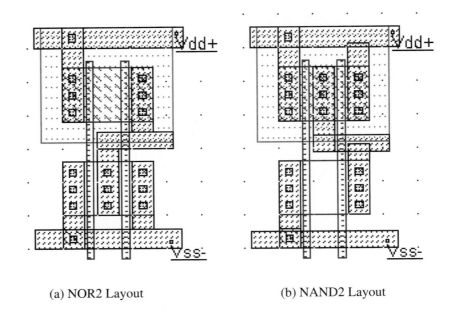

(a) NOR2 Layout   (b) NAND2 Layout

Consider first a NOR2 gate where the nFETs are in parallel. Examining the general core layout, we see that the 2-fingered nMOS transistor at this stage is already the desired result, since it is the same as two nFETs in parallel. The Metal1 lines on the left and right sides (as shown in Figure 8.24) provide the nFET connections shown earlier in Figure 8.20. To create series pFETs, extend the left Metal1 up to VDD and adjust the bottom to end at the bottom of the pdiff. Then erase (delete) the Metal1 and contacts between the two gates. Finally, extend the right Metal1 line downward, and add a Metal1 segment that touches the Metal1 line coming up from the middle of the nFETs. Adding a VDD contact to n-well and a VSS contact to p-substrate, connecting the respective polygates, gives the final result shown in Figure 8.25(a). You may want to save this gate for future reference.

The NAND2 gate in Figure 8.25(b) is obtained with the same procedures, only now Figure 8.22 shows that the nFETs are in series while the pFETs are in parallel. Starting from the original core cell in Figure 8.24, we create the series nFETs by deleting the right Metal1 down to VSS, eliminating the middle contacts and Metal1 line, and extending the right Metal1 line up. The parallel-connected pFETs are provided already in the core, so all we need to do is to extend the left and right Metal1 lines to VDD, pull down the central Metal1, and add a Metal1 output and substrate/n-well contacts. This completes the layout of the NAND2 gate.

This example illustrates the restrictions introduced by the VDD–VSS edge-to-edge spacing of 54 λ with this layout philosophy. It would be difficult to use transistors with large channel widths, since they would not fit in the allotted area without violating design rules. This limitation, however, is due to the fact that we have chosen to draw the FETs with the poly gates running vertically. If we redraw the circuits as in Figure 8.26, then we can accommodate larger widths as needed by increasing the extent X originally shown in Figure 8.8. The point here is that the FET can be rotated in either direction, so long as the

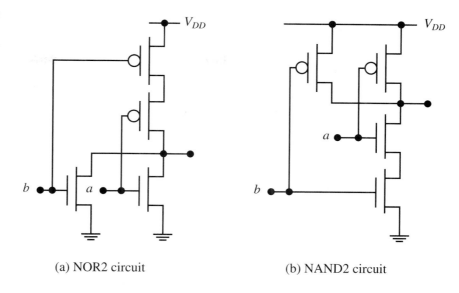

(a) NOR2 circuit   (b) NAND2 circuit

operation maintains the same topology and does not introduce too much capacitance into the output node. There isn't any rule that says that the nFETs and pFETs must have the same orientation, so we can select the layout style that works. An example is the NAND2 gate shown in Figure 8.27. This uses the same orientation for the pFETs while the nFETs are "stacked" vertically with a minimum poly-to-poly spacing for the series gates. This design uses larger transistors of both polarities. The pFETs can be rotated and separated to provide for larger values of $(W/L)_p$, if necessary.

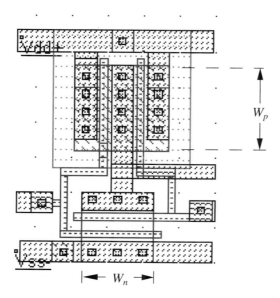

Figure 8.28:
NOT–buffer
combination

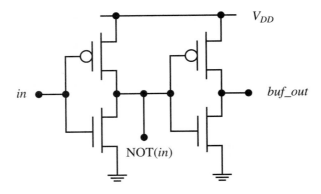

NOT–Buffer Circuit

The core circuit in Figure 8.24 can also be used to create a NOT–Buffer combination. As shown in Figure 8.28, this cell contains two cascaded inverters. The buffered output of the second stage has the same logical value as the input *in*, but is delayed in time. Buffers are used to strengthen the drive capability of a signal. The inversion operation provided by the first inverter is available, if needed. This is a very useful cell to have for general-purpose applications.

The layout for the cell is shown in Figure 8.29. This is made up of two back-to-back inverters that share common V_{DD} and V_{SS} connections to reduce the cell area. The output of the first (left) inverter is fed to the input of the second (right) inverter using a poly line underneath the Metal1 connection to V_{SS}. This also provides the inverted output NOT(*in1*)

Figure 8.29:
Layout for the
NOT–buffer cell

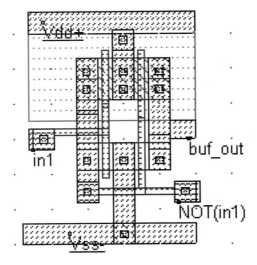

Figure 8.30:
Simulation results
for NOT–buffer
circuit

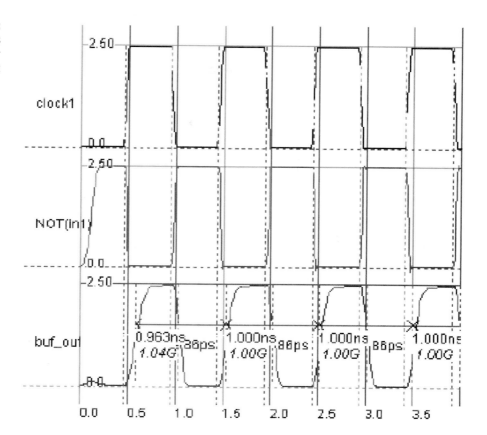

in the circuit. The SPICE simulation in Figure 8.30 illustrates the delay when driving a 40-fF capacitor at the output. The slope of the output waveform transitions can be increased by using larger FETs in the second stage. Note, however, that increasing the size of the second-stage transistors will increase the load capacitance seen by the first stage and slow it down.

3-Input Gates

The design and layout ideas can be extended to NOR3 and NAND3 gates. A 3-input NOR gate has three nFETs in parallel and three pFETs in series. The circuit diagram in Figure 8.31 provides the details of the topology. This approach again uses 3-transistor groups that are wired to create series and parallel connections.

The NAND3 gate is the dual circuit and consists of three nFETs in series and three pFETs in parallel. This leads to the circuit shown in Figure 8.32. Note that only a few wiring changes are needed to transform the NOR3 into a NAND3 circuit.

Figure 8.31:
NOR3 circuit

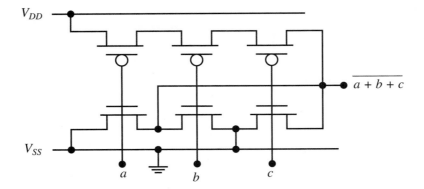

Figure 8.32:
NAND3 circuit

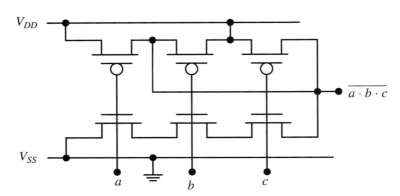

The layouts for the 3-input gates are reproduced in Figure 8.33. Both circuits have nFETs of widths 12 λ and pFETs of widths 20 λ and follow their respective circuit diagrams. It is worth mentioning that the poly-input lines need to be routed to I/O locations, with poly contacts and metal **interconnect** added to complete the cell.

We note in passing that NAND and NOR gates with four or more inputs can be designed using the same approach. In practice, however, the large fan-in[1] gates are too slow due to the existence of long series chains of FETs. High-speed design relies on carefully designing logic chains using small fan-in gates and concentrating on the FET sizing problem.

8.2.5   Simulating Logic Gates

Microwind's SPICE implementation supports a limited range of input functions, but they can be used to provide excellent simulation results for multiple-input gates.

---

1. Fan-in is the number of inputs.

Figure 8.33:
Layouts for
3-input gates

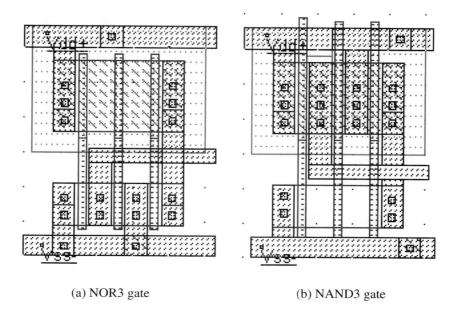

(a) NOR3 gate   (b) NAND3 gate

The easiest approach to simulating a logic circuit is to apply a Clock input to each gate input, and then to adjust the timing parameters. First select and click the polygon where you want to apply the input. When you push the **Clock** button on the **Palette** window, the dialog window in Figure 8.34 appears. Microwind uses a standard SPICE construction (with slightly different terminology) to describe the waveform. The low and high values are set at the default values of VSS (0 V) and VDD, respectively. The timing parameters are

- **Time start**—the delay from $t = 0$ before the pulse is applied
- **Rise time**—the time for the signal to rise from a low value to the high value in a linear ramp
- **Time pulse**—the pulse width, i.e., the length of time that the pulse is kept at the high value
- **Fall time**—the time interval for the pulse to fall from a high to a low value

The clock waveform is periodic, and will repeat automatically with a period T defined by the waveform segment. The values can be set manually, or the **Slower** and **Faster** buttons can be used to adjust the frequency. The default values have been selected to provide default simulations using the selected technology. Pushing the **Assign** button assigns the input to the point you specified.

The program automatically adjusts the characteristics of pulse inputs as they are applied to different inputs of the circuit. The first pulse has a period T, while the second pulse that is used will have a period of $2T$. The third applied pulse has a period of $4T$, and

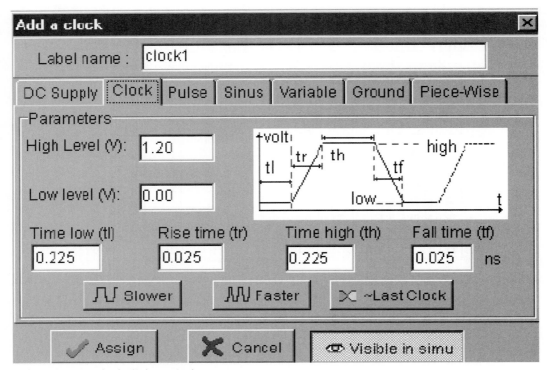

Figure 8.34: Clock dialog window

so on, with the period doubling with each added input. This staggers the inputs in time, as shown in Figure 8.35. This provides inputs that will create all of the input combinations as the simulation is run. The logic operation can thus be observed in a transient simulation while studying the switching times.

> *Remember: If you do not see the node you want on the simulation plots, then you need to make it visible using the **Navigator** button.*

Figure 8.35:
Clock sequencing

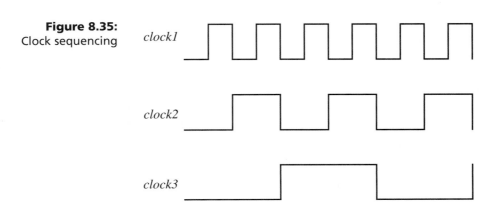

8.3 Complex Logic Gates

Complex logic gates give a combination of logical OR and AND functions in a single circuit. This CMOS feature is quite useful for merging functions and designing small circuits. An example is given by

$$f = \overline{a \cdot b + c} \qquad (8.29)$$

which provides one AND and one OR operation; the NOT is a characteristic of CMOS logic and is a consequence of the topology. To design the CMOS gate, note that $f = 0$ if either $a \cdot b = 1$ or $c = 1$. From our study of the NAND and NOR gates, we know that series-connected nFETs give the AND operation, while parallel-connected nFETs produce the OR. Thus, we construct an nMOS logic array with three nFETs. Two transistors (Mna and Mnb) will have inputs a and b and will be wired in series, and this group will be in parallel with another nFET (Mnc) that has c as the input.

The circuit is shown in Figure 8.36. The pFET wiring was designed using series-parallel nFET/pFET structuring. Since Mna and Mnb are in series, their pFET counterparts, Mpa and Mpb, must be in parallel. Using the same line of reasoning, the group (Mna, Mnb) is in parallel with Mnc, so that Mpc must be in series with the group (Mpa, Mpb). This approach ensures that the output is always well defined as either a 0 or 1 voltage. Figure 8.37 shows the layout based on the circuit diagram. The transistor widths were reduced to accommodate the wiring between Mpa and Mpb. This slows down the circuit because the resistances are larger. In addition, note that the capacitance between Mpa, Mpb, and Mpc is relatively large.

As another example, consider the four-input function

$$g = \overline{a \cdot b + c \cdot d} \qquad (8.30)$$

Figure 8.36:
Complex CMOS
logic gate example

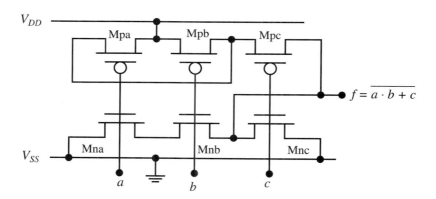

Figure 8.37:
Layout for complex
logic gate example

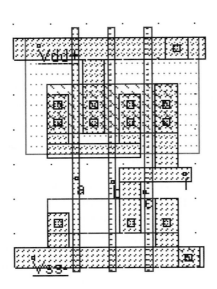

This is sometimes called an **AND-OR-INVERT 22** operation (**AOI22**), which is obtained by just following the order of operation. The CMOS circuit is shown in Figure 8.38. The structuring can be understood by associating the nFETs with the expression, then applying series–parallel arguments to wire the pFETs. Mna and Mnb are in series corresponding to the term $a \cdot b$. The same reasoning holds for Mnc and Mnd, which are in series corresponding to the term $c \cdot d$. The two groups (Mna, Mnb) and (Mnc, Mnd) are in parallel because of the **OR** operation (+) in the expression. The pFET wiring is obtained using series-parallel arguments. Mpa and Mpb are in parallel, as are Mpc and Mpd. The groups (Mpa, Mpb) and (Mpc, Mpd) are in series, since their nFET counterparts are in parallel. A layout for the AOI22 circuit is shown in Figure 8.39. This uses a Metal2 line to wire Mpc and Mpd in parallel and still allow the output to be routed on Metal1. This is our first example of how multiple metal layers can help solve wiring problems.

Figure 8.38:
AOI22 circuit

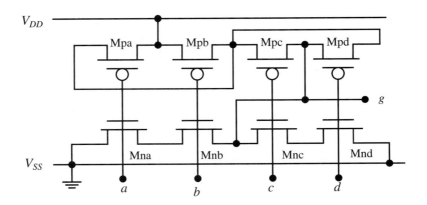

Figure 8.39:
AOI22 layout using
Metal1 and Metal2

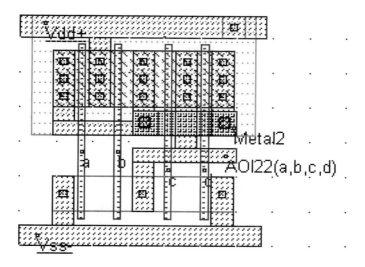

Figure 8.39:
AOI22 layout using
Metal1 and Metal2

A related function is the **OR-AND-INVERT 22 (OAI22)** circuit shown in Figure 8.40. This implements the function

$$h = \overline{(a+b) \cdot (c+d)} \tag{8.31}$$

and is the dual of the AOI22 circuit. The layout can be deduced from the AOI22 circuit using series-parallel nFET/pFET arranging, and it is left as an exercise for the reader.

These examples illustrate that it is relatively easy to design complex logic gates in CMOS using series-parallel structuring. It is not always possible to create a layout that uses strings of nFETs or pFETs. To determine if single ndiff and pdiff regions can be used, see if you can trace a closed, continuous path that passes through every FET in the circuit, but only one pass per FET is permitted. Multiple passes through a node are acceptable. If you can draw the line without lifting the pencil, then the resulting trace is called an **Euler**

Figure 8.40:
OAI22 logic gate

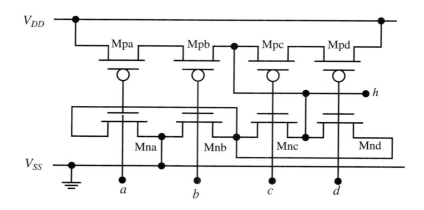

Figure 8.41:
Example of an
Euler graph trace

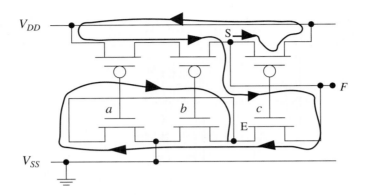

graph, and it is possible to use single ndiff and pdiff regions. If you cannot construct an Euler Graph, groups of transistors need to be designed and then wired together. An example of an Euler Graph for the function

$$F = \overline{(a + b) \cdot c} \tag{8.32}$$

is shown in Figure 8.41. The point S denotes the start of the trace, and E is where the loop ends. Since the path closes, the layout can use shared diff regions.

It is important to remember that complex logic gates are useful because a single circuit can provide multiple AND and OR operations. This provides a small-area option to the classical approach of using cascades of primitive gates to build logic functions. The drawback is that the gates may be slow because of the series-connected transistors and the associated resistances and capacitances. In the practical sense, however, not every gate needs to be fast, since the maximum system speed is established by a few critical paths.

● ● ● ● ● ● ● ● ● ● ● ● ● ● ● ● ●

8.4   The Microwind Compile Command

Microwind is able to create the layout of a CMOS logic circuit from a one-line logical expression. From the Microwind Main Menu bar, execute the command sequence

Compile $\Rightarrow$
    Compile one Line

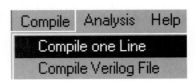

Figure 8.42:
Compiler one-line
input

as shown. This opens up the dialog window shown in Figure 8.42. The desired expression is then entered into the space provided using the proper syntax for the logical operations. In the example shown, the function is

$$
\begin{aligned}
f &= A \,\&\, (B|C) \\
  &= A \cdot (B + C)
\end{aligned}
$$

(8.33)

i.e., & is the AND operation, and | is the OR. The parenthesis imply precedence of the OR operation. Clicking on the **Compile** button creates the layout plot shown in Figure 8.43. The layout has been annotated to show the formation of the logic. The first gate forms the inverted function

$$
\bar{f} = \overline{A \cdot (B + C)}
$$

(8.34)

Figure 8.43:
Complex logic
gate created by
the **Compile**
operation

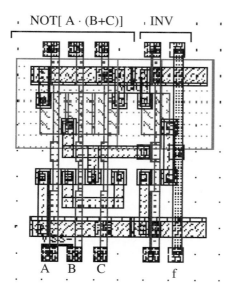

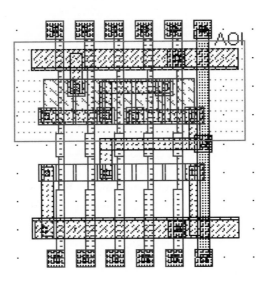

because CMOS logic gates automatically provide inversion. This is then fed through an inverter to obtain the desired result.

The **Compile** function can also provide the inversion of a function using the tilde (~) after the = sign. The syntax is

$$\text{Output} \;=\sim\; (\text{Expression}) \tag{8.35}$$

such that the combination =~ does not have a space between the characters. An example is given by

$$\text{AOI} =\sim (\text{inA \& inB})|(a \;\&\; b \;\&\; c) \tag{8.36}$$

This compiles to the layout shown in Figure 8.44. You can check the nFET array to see there are two nFETs in series on the left side, and that this group is in parallel with a group of three series-connected nFETs on the right side. The **NOT** operation cannot be applied to input variables. If the expression cannot be compiled, usually because of a syntax problem, an error notice will appear on the dialog line at the bottom of the work area.

The **Compile** function in Microwind is a useful tool for both learning and producing CMOS circuits quickly. You should try several different inputs and study the results, as this is an excellent learning tool. It is noted that, although the compiled cells should be functional, the layouts are not always the most compact designs possible. There will be instances when a hand-drawn layout will be a more compact or faster (or both) solution.

This should not be considered a fault of the Microwind program. Design automation tools have not yet evolved to the point where computer algorithms can create the most efficient solutions.

If this capability has captured your interest, then you should read Chapters 15 and 16, which deal with automated design and circuit compilation using the Dsch program.

● ● ● ● ● ● ● ● ● ● ● ● ● ● · ·

8.5   Tri-State Circuits

A standard logic gate has outputs of 0 and 1. In a tri-state circuit, the output can also be in a **Hi-Z** (high impedance) state, giving the circuit three distinct states.

The simplest technique for implementing a tri-state inverter is to add two additional transistors, M1 and M2, to the basic NOT gate, as in Figure 8.45. The nFET M2 is controlled by an enable signal, *En*, while $\overline{En}$ is applied to the gate of pFET M2. The operation is straightforward. If *En* = 1, then both M1 and M2 are ON, and the circuit functions as a normal inverter with $f(a) = \overline{a}$. Changing the control signal to *En* = 0 turns off both M1 and M2. The output "floats" since it sees two open switches. This defines the Hi-Z or "tri-state" condition where the inputs have no effect on the output. Physically, the parasitic capacitance can hold the voltage for awhile, unless the output is connected to another circuit that changes the voltage.

The layout is shown in Figure 8.46. It consists of two groups of series-connected FETs and follows the circuit diagram. There is an implied inverter somewhere that takes *En* and inverts it to $\overline{En}$. Reversing the enable signals creates a circuit that tri-states when *En* = 1.

Figure 8.45:
Tri-stated inverter

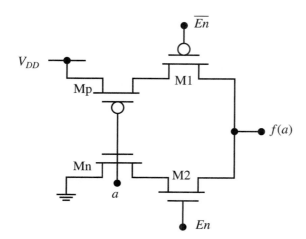

Figure 8.46:
Tri-state circuit
layout

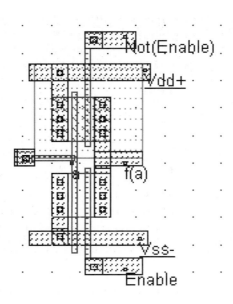

8.6   Large FETs

Sometimes, it is necessary to design transistors with large channel widths. This usually occurs when the devices must accommodate large-current flow levels. A few layout problems arise if W is large compared to the channel length, L, and these are worth discussing before proceeding further.

The first issue is the number of active contacts used to connect ndiff or pdiff regions to metal. In general, one should always use as many contacts as possible, subject to the design rule limits. This originates from two main considerations: attaining proper current flow through a transistor, and minimizing the effect of parasitic resistance.

The current flow pattern through a MOSFET is established by the location of the drain and source contacts as dictated by the laws of electrostatics. FETs conduct electricity using **drift current**; this type of current is from electrical charges moving under the influence of an electric field. Even though the diff and pdiff regions are heavily doped, they are not ideal conductors. This means that the potential (voltage) on a doped region is not a constant, but varies with location. In simple terms, current enters the device through contacts on one side and exits through the contacts on the other. Consider the small transistor shown in Figure 8.47(a). Since W is small, the current flow lines spread out and pass through most of the width. If we increase W without changing the number of contacts, as in Figure 8.47(b), then the flow is concentrated on the top of the device. The electrical W will be smaller than the geometrical value because the current does not spread out. Adding an extra contact on both sides spreads the current across the device, as in Figure 8.47(c).

Figure 8.47:
Current flow lines
in a FET

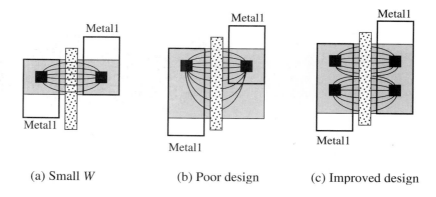

(a) Small *W*　　　　(b) Poor design　　　　(c) Improved design

Another reason that we add contacts is because every contact has a **contact resistance** that adds to the total resistance in the line. Figure 8.48 provides a cross-sectional view of a Metal1–Contact–ndiff conduction path. Each section of the chip has resistance. In the drawing, R_m is the Metal1 resistance, R_c is the contact resistance, and R_n is the ndiff resistance. The equivalent circuit shows that the contributions are in series, so that the total resistance from the metal input to the FET is

$$R = R_m + R_c + R_n \tag{8.37}$$

If we replace the single contact by a group of contacts, then the effect of R_c is reduced. Figure 8.49 shows an example using five contacts. The equivalent circuit shows that all R_c contributions are in parallel, which reduces the resistance from point A to ndiff to $(R_c/5)$.

Figure 8.48:
Contact resistance

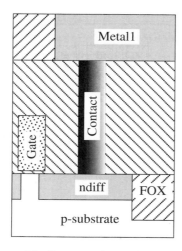

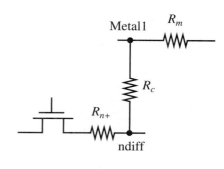

(a) Cross-sectional view　　　　(b) Equivalent circuit

Figure 8.49:
Multiple contacts

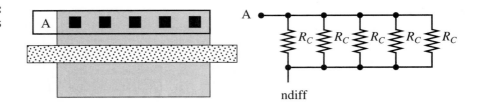

ndiff

In general, using M contacts gives an reduction to (R_c/M), so that we usually add as many as possible.

In certain cases, transistors with very large widths, W, are required. If we use a single gate, then the FET layout may get elongated and not fit well into the circuit. Some electrical problems also arise. Usually we use multiple-gate fingers to design more compact devices. The idea is shown in Figure 8.50. A wide FET with large width is shown in Figure 8.50(a). This can be "folded" to the layout shown in Figure 8.50(b), which uses two FETs having a width of $(W/2)$. The equivalent circuit illustrates the fact that the transistors are in parallel with a common gate G. The number of multiple fingers can be increased for very wide transistors. Note that B has a smaller capacitance than A because of reduced junction-capacitance contributions.

Recall that we used the Microwind MOS Layout Generator to create individual transistors that share ndiff or pdiff regions. The MOS Layout Generator produces the basic transistor structure according to the width (for each finger) and the number of fingers. Gate connections must be completed manually. An example is shown in Figure 8.51. The width of each finger was entered as 50 λ (= 5 μm in the 0.18 μm technology). The Metal1 lines were connected to create four parallel transistors, and the gate poly lines have been combined into one. Dual gate contacts were created on both sides for symmetric driving.

Figure 8.50:
Multiple gate
finger layout

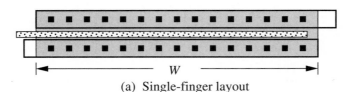

W

(a) Single-finger layout

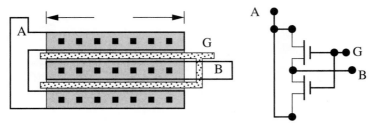

(b) Equivalent parallel devices

Figure 8.51:
Parallel FETs
using multiple
gate fingers

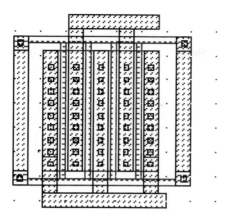

The structure is equivalent to a single nFET with a total width of $4 \times 50 \, \lambda = 200 \, \lambda$. Since the channel length of each FET is $2 \, \lambda$, the aspect ratio of the entire device is equal to $(W/L) = 100$. In digital designs, this would be considered an extremely large transistor and would only be found in high-current driver circuits. On the other hand, many analog CMOS circuits are designed using large transistors, making multiple-finger parallel devices more common in those applications. The layout of large transistors for analog circuits is discussed in Chapter 18.

8.7   Transmission Gates and Pass Logic

Every logic circuit examined thus far is based on the use of complementary nFET/pFET pairs. We generally refer to this as a "standard" or "full complementary" design. This is not the only approach that can be used. CMOS is rich in design styles, each with its own characteristics.

A classical technique is based on the use of **transmission gates** (**TG**s). A TG is a parallel-connected nFET/pFET pair that acts as a logic-controlled switch. The circuit is illustrated in Figure 8.52(a). The nFET is controlled by a **logic signal**, s, while the pFET gate is wired to the complement, $\bar{s}$. When $s = 1$, both FETs are ON, and the input a is transmitted to the output so that $f = a$. The TG is capable of full-rail transmission since the nFET can pass a strong 0 (0 V) and the pFET can pass a strong 1 (V_{DD}). If $s = 0$, then both transistors are OFF and the output is in a Hi-Z state. In this case, f is undefined. The operation is summarized in Figure 8.52(b). A simplified symbol for the TG is given in Figure 8.52(c).

Layout of a TG is straightforward, since it consists of only two transistors. Note that there are no direct connections to the power supply or ground, but one should bias the n-

Figure 8.52:
Transmission gate

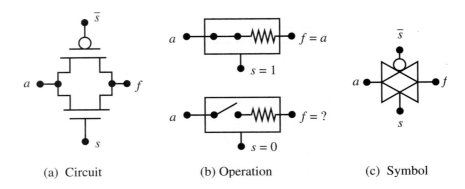

(a) Circuit   (b) Operation   (c) Symbol

well and p-substrate. Figure 8.53 shows the basic layout. To illustrate the electrical characteristics, V_{DD} has been applied to the gate of the nFET, while the pFET gate has been
grounded. A full-rail clocking signal that swings from 0 to 2.0 V has been used as a source
on the left side, and a 40 fF (0.04 pF) capacitor has been added to the right side. The simulation result is shown in Figure 8.54. It is seen that the output waveform does indeed have
a full-rail voltage swing. The rise and fall times depend upon the parasitic resistance and
capacitance at the output node. The overall response is also determined by the input-driv-
ing circuitry that replaces the clock in a more realistic application.

Transmission gates can be used to build simple logic circuits. An example is the 2-to-1
multiplexor in Figure 8.55(a). Two input data paths, d_0 and d_1, are input into the circuit. A
select signal, s, determines which data path is sent to the output. If $s = 0$, then TG0 is
closed while TG1 is open; this gives an output of $f = d_0$. If $s = 1$, the situation is reversed
with TG0 open and TG1 closed, so that $f = d_1$. Figure 8.55(b) shows one approach to the
layout of the TG MUX. This shares the output line (f) to reduce the area.

Figure 8.53:
Basic TG layout

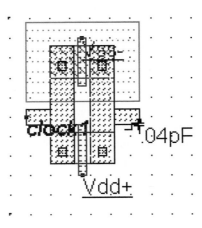

Figure 8.54:
Pass characteristics
of a transmission
gate

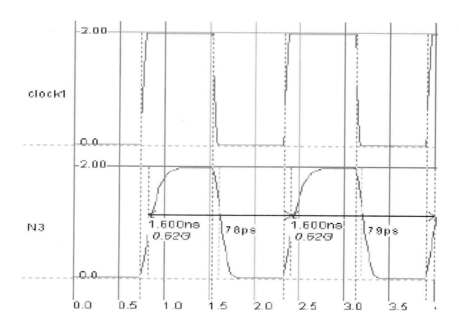

Figure 8.55:
Transmission gate
2:1 MUX

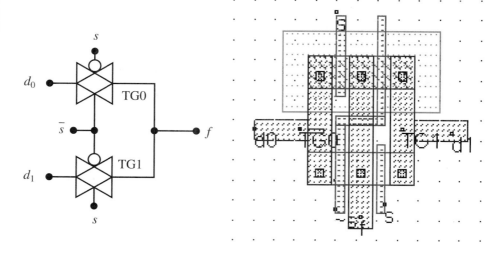

(a) TG circuit   (b) Layout example

Figure 8.56:
Exclusive-OR
logic symbol
and truth table

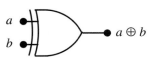

$a \oplus b$

(a) Shape-specific

a	b	$a \oplus b$
0	0	0
0	1	1
1	0	1
1	1	0

(b) Truth table

A useful variation of the 2:1 MUX is the **exclusive-OR (XOR)** gate. The logic symbol and truth table are shown in Figure 8.56. A TG XOR circuit is drawn in Figure 8.57. The circuit feeds a to TG0 and uses a NOT gate to provide $\bar{a}$ as an input to TG1. When $b = 0$, TG0 is closed and the output is a; in Boolean terms, this case is represented by $a \cdot \bar{b}$. For the case $b = 1$, TG1 is closed and the output is $\bar{a}$, which is equivalent to $\bar{a} \cdot b$. Since the output is one OR the other, we note the identity

$$a \oplus b = a \cdot \bar{b} + \bar{a} \cdot b \qquad (8.38)$$

to conclude that this is the XOR operation. An **XNOR (exclusive-NOR)** gate can be obtained by simply interchanging the b and $\bar{b}$ signals:

$$a \cdot b + \bar{a} \cdot \bar{b} = \overline{a \oplus b} \qquad (8.39)$$

TG circuits of this type are interesting because they are based on simple switch-logic principles.

The TG has the same logical switching characteristics as an nFET, so that we can also build multiplexors and XOR/XNOR circuits using only nFETs. The main drawback is the

Figure 8.57:
TG-based XOR
circuit

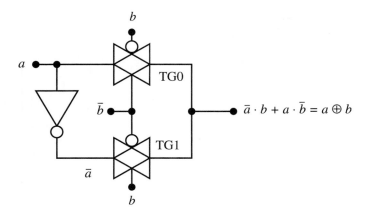

threshold voltage loss when a logic 1 high voltage is applied to the input of an nFET used as a pass transistor. This may or may not be a major problem in the application.

● ● ● ● ● ● ● ● ● ● ● ● ● ● ● ● ·

8.8 References

CMOS logic design is a standard topic that is covered in many books. Here is a short listing.

[8.1] Martin, K., *Digital Integrated Circuit Design*. New York: Oxford University Press, 2000.

[8.2] Uyemura, J. P., *A First Course in Digital Systems Design*. Pacific Grove, CA: Brooks-Cole, 2000.

[8.3] Uyemura, J. P., *CMOS Logic Circuit Design*. Norwell, MA: Kluwer Academic Publishers, 1999.

[8.4] Uyemura, J. P., *Introduction to VLSI Circuits and Systems*. New York: John Wiley & Sons, 2002.

[8.5] Wolf, W. H., *Modern VLSI Design, Third Edition*. Upper Saddle River, NJ: Prentice Hall, 2002.

● ● ● ● ● ● ● ● ● ● ● ● ● ● ● ● ·

8.9 Exercises

8.1 Create the layout for an inverter with $(W/L)_n = 4 = (W/L)_p$. Simulate the DC and transient characteristics when a 0.025 pF capacitor is used as a load.

8.2 Design a symmetric inverter assuming $r = 3$. Use $(W/L)_n = 4$ as a basis. Obtain the VTC and see if the circuit is symmetric with your technology choice.

8.3 Design a CMOS gate that implements the function

$$f = \overline{a \cdot b + a \cdot c + a \cdot d} \qquad (8.40)$$

using the smallest number of FETs. Simulate to verify the logic.

8.4 Use the MOS Layout Generator to design a multiple-finger nFET with an aspect ratio of 100. Try to obtain a square overall shape.

CHAPTER

9

Standard Cell Design— Layouts and Wiring

High-density digital VLSI systems are created from primitive components using the concepts of repetition and structural regularity. Standard cell design is an approach that uses a collection of logic cells to create more complex networks. In this chapter, we investigate some of the details needed to use standard cells in a layout.

9.1 Cell Hierarchies

All complex logic functions are implemented using cascades of primitive logic gates. VLSI is based on the premise that there are a finite number of primitive gates that are needed to build a system. Once we design a CMOS circuit, including the layout and electrical characterization, it can be stored in a **cell library**, and copies of it can be used as needed. This simple observation makes the design of complex systems tangible: it reduces the complexity of the problem by restricting the design to using a finite set of basic components. It forms the basis of hierarchical design.

Since this may sound somewhat abstract, let us use the example of a **full adder (FA)**. The block symbol and function table are shown in Figure 9.1. With inputs of a_n, b_n, and the **carry-in bit**, c_n, the outputs are the sum,

$$s_n = a_n \oplus b_n \oplus c_n \qquad (9.1)$$

and carry-out equation,

$$c_{n+1} = a_n \cdot b_n + c_n \cdot (a_n \oplus b_n) \qquad (9.2)$$

in one of their most familiar forms. The repetition of primitive logic functions is quite obvious. The sum bit uses two XORs, and $(a_n \oplus b_n)$ also appears in the carry-out equation.

Figure 9.1:
Full-adder symbol
and function table

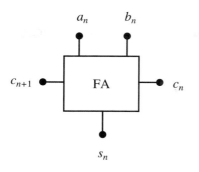

a_n	b_n	c_n	s_n	c_{n+1}
0	0	0	0	0
0	1	0	1	0
1	0	0	1	0
1	1	0	0	1
0	0	1	1	0
0	1	1	0	1
1	0	1	0	1
1	1	1	1	0

(a) Block symbol   (b) Function table

The AND and OR are also used to compute c_{n+1}. A basic CMOS implementation of a full adder can be built using three basic cells (AND, OR, XOR) that are reproduced as needed and wired together in the proper manner. Note that when $c_n = 0$, the circuit degenerates to a half adder (HA) that takes a and b and produces the sum, s, and carry, c, from

$$s = a \oplus b$$
$$c = a \cdot b$$

(9.3)

This function only requires two operations (AND, XOR) in the library.

9.1.1   Design of an XOR Cell

The AND is considered a primitive operation, but the XOR is given by

$$a \oplus b = a \cdot \bar{b} + \bar{a} \cdot b$$

(9.4)

A simple, static XOR circuit can be built by using the DeMorgan theorems to write the expression in the form

$$a \oplus b = \overline{\overline{a \cdot b} + \overline{\bar{a} \cdot \bar{b}}}$$
$$= \overline{[a \cdot b + (\overline{a + b})]}$$

(9.5)

The logic diagram for this implementation is shown in Figure 9.2. It consists of a NOR2 and an AOI21 gate that use the same inputs, (a,b).

This composite cell can be created using the **Microwind Compile One Line** operation. First, we create the NOR2 cell by entering the expression:

$$NOR2 =\sim (A|B)$$

(9.6)

Figure 9.2:
Exclusive-OR
(XOR) logic

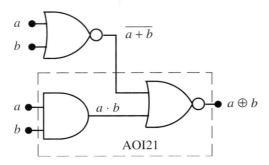

Clicking on the **Compile** button generates the layout shown in Figure 9.3. This is the familiar NOR2 cell discussed Section 8.4. We will perform a **Save As** operation on the layout so that we can access it later. Next, we will add the AOI22 cell to the layout by compiling the logical expression

$$AOI21 =\sim In|(A\&B) \tag{9.7}$$

The order has been chosen to place the IN input close to the output of the NOR2 layout. Clicking the **Compile** button generates the AOI21 cell and automatically aligns it with the VDD and VSS lines, as seen in Figure 9.4. Maintaining a constant VDD–VSS pitch value for every cell (60 λ in our examples) allows us to create cell groups that interface with one another. Since the layout now has all of the needed logic gates, we simply connect the inputs and outputs to produce the XOR(A,B) function. First, the output of the NOR2 gate is connected to IN of the AOI21 gate using a Metal1 line. Then, we route A and B inputs horizontally below VSS with Metal3 lines and connect them to the A and B gate inputs using vertical Metal2 lines. The output is also taken out at Metal3, although it already is available from the vertical Metal2 line. This gives the final layout shown in Figure 9.5. We will save this cell as an XOR2 layout.

Figure 9.3:
NOR2(A,B) cell
created with the
Microwind Compile
window

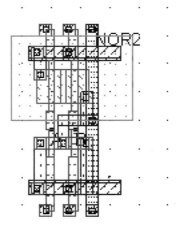

Figure 9.4:
Layout after
generation of
AOI21 cell

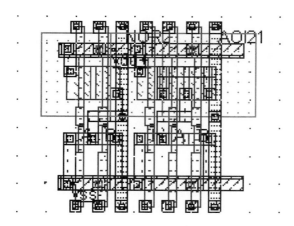

The Contact Layout Generator feature is used to create Metal2-to-Metal3 vias; Figure 9.6 shows the location on the Palette window. Activating this button opens the dialog window shown in Figure 9.7. This allows the user to select the specific contacts that are desired. In deep submicron designs, contacts and vias can be stacked on top of each other to increase the packing density. Creating the contact in this manner is convenient, especially when several metal layers are used. The result can be checked with the 2-dimensional cross-sectional viewer, and the electrical contact can be verified with the View Node feature in the Navigator window. When the Navigator window is open, clicking on

Figure 9.5:
Completed
XOR(A,B) cell

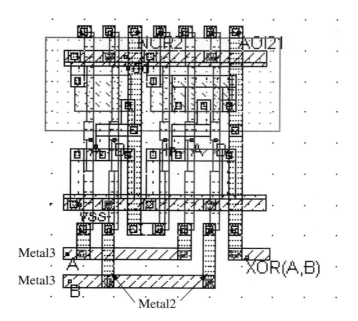

Figure 9.6:
Contact buttons on
the Palette window

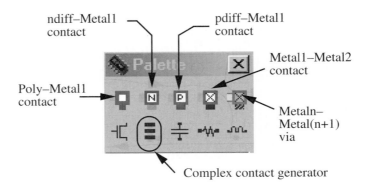

ndiff–Metal1 contact

pdiff–Metal1 contact

Metal1–Metal2 contact

Metaln– Metal(n+1) via

Poly–Metal1 contact

Complex contact generator

any point automatically will highlight all sections of the layout that are in electrical contact with the selected region. Figure 9.8 shows a test of the B input node on the layout. Note how all other sections are outlined. This is a very useful tool for checking your layouts, especially when they get complex. One key to successful design is to verify, re-verify, and check it over and over again!

9.1.2   Half-Adder Cell

Now that we have an XOR, let us build a half-adder (HA) circuit. From Equation (9.3), we see that we need an AND gate to calculate the carry bit c. Although we could design one from scratch, we will opt to again use the Compile feature by entering the expression

$$\text{and} = A \ \& \ B \tag{9.8}$$

Figure 9.7:
Contact Generator
dialog window

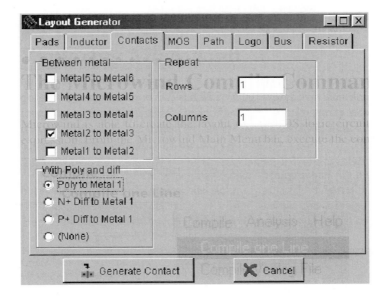

Figure 9.8:
Using the See
Navigator feature
to check electrical
continuity

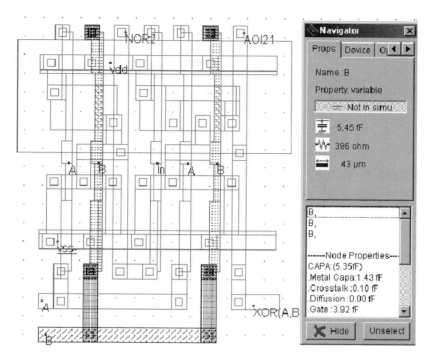

This generates the layout shown in Figure 9.9(a). Note that the circuit is a NAND2 gate followed by an inverter. This is necessary because CMOS is an inverting logic family, so that the AND2 function cannot be created directly. Since the inputs are on the left side of the cell, we will perform a **Flip Horizontal** operation to obtain the layout shown in Figure 9.9(b). This will make it easier to interface to the XOR circuit. We will save this cell as the AND2 layout.

Figure 9.9:
AND2 layout

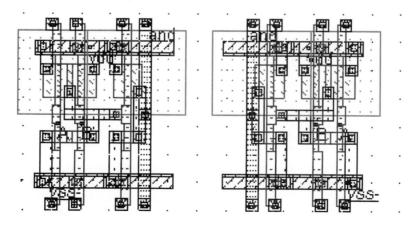

(a) Generated AND2 cell   (b) After horizontal flip

Figure 9.10:
Insert layout
command

The next operation allows us to merge two separate cell files together to obtain both the sum and the carry circuits for the half adder. We will start with the flipped AND2 cell displayed in the work area; this provides the carry bit

$$c = a \cdot b \tag{9.9}$$

for the half adder. To add the sum circuit needed in

$$s = a \oplus b \tag{9.10}$$

we go to the Microwind Menu bar and execute the command sequence

File $\Rightarrow$

Insert layout

as shown in Figure 9.10. This opens up a window that displays the contents of the Microwind folder. Find the XOR2 file that was designed in the previous example and click on it. This copies the layout into the present work area. Using the Edit commands, group and move the entire XOR2 circuit to the right of the AND2 layout until the VDD and VSS segments align. The easiest technique is to use the Move Step-by-Step feature.

Once the two cells are aligned horizontally, extend the Metal3 A and B input lines to the left. Then connect them to the proper A and B inputs of the AND2 cell. Your completed layout will look something like that shown in Figure 9.11. This is, in fact, a completed half-adder circuit! The labels in the figure have been changed to reflect their new status. The output of the AND2 gate on the left is the carry bit, c, while the output of the XOR gate on the right is the sum bit, s. Save this cell as a half adder for future use.

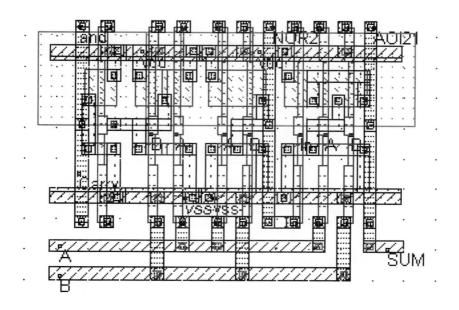

9.1.3 Full Adder

A full adder can be constructed using two half adders, as shown in Figure 9.12. This design uses two HA cells and one OR2. The layout for the full adder is shown in Figure 9.13; the cell boundaries for the HA and OR2 sections have been added to help visualize the construction. The inputs a_n, b_n, and c_n appear at the bottom I/O signal lines. The sum, s_n, is on the right side while the carry out, c_{n+1}, is on the left. Of course, the outputs can be routed as desired by adding more interconnect wiring.

The important aspect of this design is the step-by-step manner that was used to create it. The XOR was built from primitive cells and then was itself used to design a half adder. The HA cell then became the basis for the FA network. This shows how hierarchical design operates: each layer of complexity builds on simpler ones.

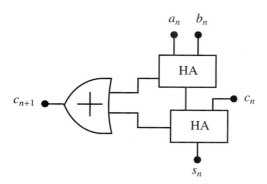

Figure 9.13:
Full-adder layout
using HA cells

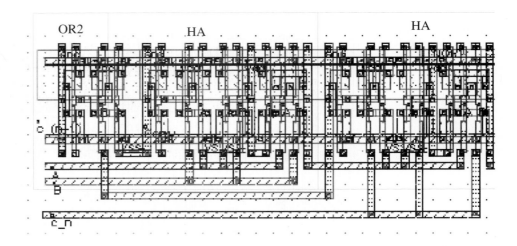

9.1.4 Cell Levels

Let us examine the complexity of each cell in the evolution of the full adder. The levels can be illustrated using the idea portrayed in Figure 9.14, where the primitive cells are at the top and the complexity increases downward. Primitive cells (NOR2, NOT, and NAND2) are used to create the "simple" cells for the XOR and AND operation. These combine to create the half-adder cell, which is the next level down and labeled as the "moderate complexity" cell level. The half adder and the OR gates are then used to create the full adder, which is labeled as being at a "higher complexity" level. This is a basic map on how to build up a large VLSI system. We mix cells at different levels of complexity to create larger cells. These then act as building blocks for even larger circuits, and the process continues until the desired functions are implemented.

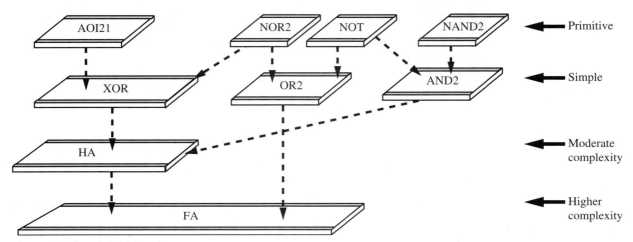

Figure 9.14: Cell hierarchy

In some cases, the existing cells will not give the desired characteristics, so we must design new ones. This is called **custom design**. It is quite complicated and time consuming and usually is used only for critical data paths or logic sections where it absolutely is necessary. If a chip is engineered from scratch without the use of a cell library, then it is called a **full custom** design. Owing to the complexity of modern digital systems, full custom designs are only found in very specialized circumstances.

● ● ● ● ● ● ● ● ● ● ● ● ● ●●●●

9.2 Cell Libraries

The full-adder example illustrates some of the basic ideas of CMOS logic design using standard cells. The XOR gate was constructed using two simpler cells, and the cell was finished by adding interconnect wiring. The three cells created in this example (NOR2, AOI21, and XOR) can be stored and used in future designs. Cell **reuse** allows us to progress without replicating what is already available. The collection of cell files is called the cell library. The larger the library, the more diverse the possible application group. Most libraries contain several cells that perform the same function but have different electrical characteristics or layout dimensions. For example, inverters with different FET sizes have different input capacitances and drive strengths; no single NOT gate will satisfy every application. This helps the designer meet the specifications of the design while battling restraints and limitations that may be imposed by layout problems.

In general, a cell library is made up of both primitive functions and larger macro functions, like adders and memories, that form the basis of the design. Every cell is documented by a schematic, a function table, an optional symbol, specific electrical characteristics, and layout dimensions with input and output port locations. Library cells should not be modified by the chip designer, as changing even one characteristic in a cell may affect others using the library. Access usually is restricted to a specialized group that is responsible for maintaining and updating the library contents.

An important characteristic of a cell library is that of uniformity. Every cell must be designed with compatible geometrical features to allow interfacing at the physical level. One obvious point of consistency is the VDD–VSS spacing, but other considerations such as routing and power distribution, are also important. Since we already have seen how the cell geometry is important, let us turn our attention to the wiring technique used in the examples thus far.

9.2.1 Port Placement and Wiring Strategies

Every cell has input and output (I/O) pads. The location of the cells is related intimately to the method used to construct cell cascades and provide interconnect wiring. Figure 9.15 illustrates two methods for placing I/O ports. In Figure 9.15(a), the ports are placed above VDD and below VSS. Interior ports (between VDD and VSS) are used in Figure 9.15(b).

Figure 9.15:
I/O port locations
for a cell

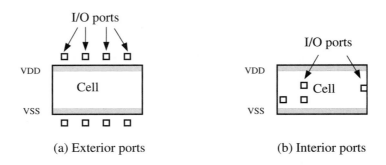

(a) Exterior ports (b) Interior ports

Although one can create a cell using both exterior and interior ports for maximum flexibility, the method used to create cell arrays usually is oriented to one or the other.

The technique used in the adder examples employs routing (or wiring) channels. Figure 9.16 illustrates the general idea. Rows of logic cells are spaced by routing-channel regions that are used to wire input and output ports. Exterior I/O ports extend into the routing channels for easy access. In this illustration, horizontal wires are created using Metal3 lines that are parallel to the VDD and VSS rails. Connections between the Metal3 lines and the I/O ports are obtained using Metal2 paths that are perpendicular to Metal3. Routing channels give reasonably straightforward wiring solutions, since interconnected wiring can be inserted as needed. It does decrease the logic density because of the spacing allocation between the rows of logic cells. However, this approach is found commonly in standard cell design because of its inherent ability to provide solutions using CAD tools.

Figure 9.16:
Routing channels

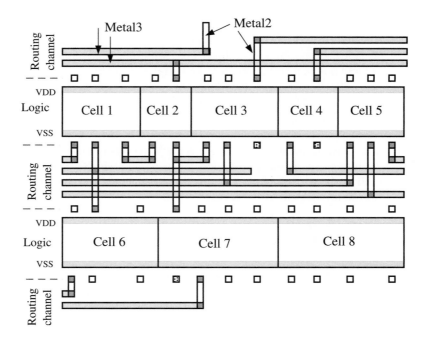

Figure 9.17:
Weinberger array

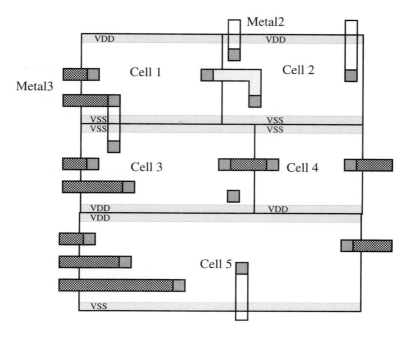

Figure 9.17:
Weinberger array

An alternate approach is to use internal I/O pads and do all of the wiring over the cells using higher-level metals. Figure 9.17 illustrates what is known as a **Weinberger Array**. Rows of cells are interfaced together by performing a vertical flip on every other row. In the drawing, this gives "upright" cells (VDD on top) on the first row, and "inverted" cells (VDD at the bottom) on the second row, and so on. Merging VDD and VSS lines together using this type of imaging allows two rows of cells to share one of the power rails. Of course, interior I/O pads must be used, since exterior pads would end up in other cells. Wiring is accomplished using higher-level metals and vias as needed. This technique can be very difficult if only one or two metal layers are available, as in older processes. However, since modern CMOS provides five or more metal layers for interconnect wiring, the approach is more attractive. Problems such as port location and via size and placement become important issues.

9.2.2   Layer-to-Layer Crosstalk

Different levels of metal interconnect are stacked according to the process flow. Capacitive coupling between successive conducting layers (such as Metal1 and Metal2) can cause unwanted signal transferal from one line to the other. This is called **crosstalk**, and it can induce incorrect voltages and logic errors. Although capacitive coupling is intrinsic in the structure, the level of coupling is determined by the layout. It is therefore an important consideration to the physical designer.

Figure 9.18:
Layer-to-layer
capacitance

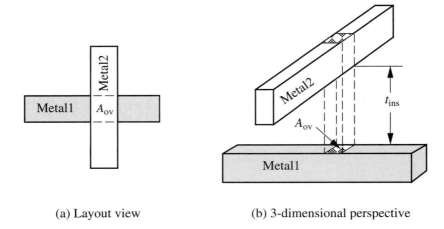

(a) Layout view (b) 3-dimensional perspective

Consider a region where lines of Metal1 and Metal2 overlap, as illustrated in Figure 9.18. The top view in Figure 9.18(a) translates to the physical structure in Figure 9.18(b). The parallel-plate capacitor formula can be used to estimate the capacitance as

$$C_{M1-M2} = \frac{\varepsilon_{ins} A_{ov}}{t_{ins}} \tag{9.11}$$

where t_{ins} is the thickness of the insulator, ε_{ins} is the dielectric permittivity, and A_{ov} is the overlap area. In processing specifications, this is usually described by the capacitance per unit area

$$\frac{C_{M1-M2}}{A_{ov}} = \frac{\varepsilon_{ins}}{t_{ins}} \tag{9.12}$$

since the layout of the lines determines the actual value of A_{ov}. For example, in the *cmos018.rul* Microwind file, the Metal1–Metal2 capacitance is cm2me = 50 aF/μm^2, where 1 aF = 1 attofarad = 10^{-18} F. To minimize the Metal1–Metal2 coupling capacitance, we want to keep A_{ov} as small as possible. This brings up a general rule in layout design

> *Lines on alternating conducting layers should be perpendicular to each other.*

This means that Metal1 lines should be perpendicular to Metal2 lines, Metal2 lines should be perpendicular to Metal3 lines, and so on. Although routing requirements do not always permit one to follow the rule, the general philosophy should be maintained. Figure 9.19 shows this from the layout perspective. Other interconnect problems are discussed in Chapter 12.

Figure 9.19:
General orientation
of alternating
conducting lines

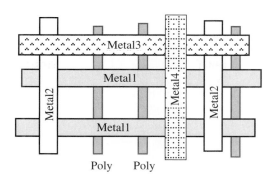

9.2.3   Scaled Driver Chains

High-speed CMOS cascades are designed by carefully selecting the size for the transistors in every stage. A well known problem in VLSI design is driving a large capacitance, C_L, that is larger than a "normal" input capacitance of a logic gate. An analysis shows that the delay is minimized by using a chain of inverters with increasing FET sizes, as illustrated in Figure 9.20. Given the input capacitance, C_1, into the first stage, the number of stages, N, needed to minimize the delay from the input of Stage 1 to C_L is estimated from [9.6]

$$N = \ln\left(\frac{C_L}{C_1}\right)$$
(9.13)

by selecting the nearest integer (even for non-inverting, odd for inverting). The inverter sizes increase from input to the load such that

$$\left(\frac{W}{L}\right)_2 = S\left(\frac{W}{L}\right)_1$$
$$\left(\frac{W}{L}\right)_3 = S\left(\frac{W}{L}\right)_2 = S^2\left(\frac{W}{L}\right)_1$$
(9.14)

Figure 9.20:
Driving a large
capacitor with a
scaled cascade

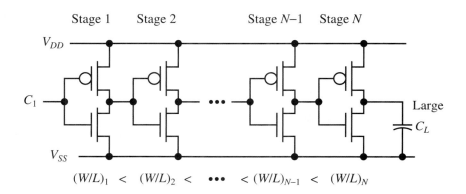

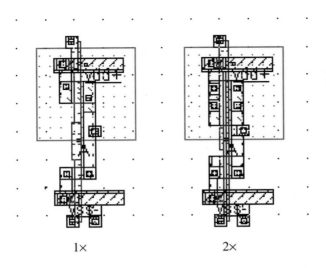

1×   2×

and so on, where $S > 1$ is a process dependent scaling factor. Typical values range from around 3 to about 6.

To accommodate this and other design problems, it is common to include scaled gates in a library. If we denote the smallest gate as a 1× design, then we might include 2×, 3×, 5×, and other sizes in the collection. This notation means that if a 1× has a FET width of W_1, the an m× gate uses a FET with a width of $W_m = mW_1$. Figure 9.21 shows an example of 1× and 2× inverter cells. To create 3× and higher designs, multiple-fingered devices are needed. It should be noted that the parasitic resistance and capacitances change with the size. If a 1× gate has a resistance of R_1 and an input capacitance of C_1, then an m× scaled gate has corresponding values of

$$R_m = \frac{R_1}{m} \qquad C_m = mC_1 \qquad (9.15)$$

This is important for the electrical design.

The concept of a scaled inverter chain can be extended to arbitrary-logic gates using a technique known as **logical effort**. In this approach, every gate is characterized by the ratio

$$g = \frac{C_{in}}{C_{ref}} \qquad (9.16)$$

where C_{in} is the input capacitance of the gate, and C_{ref} is the input capacitance to a reference inverter. Here g is called the **logical effort parameter**. The parameter

$$h = \frac{C_{out}}{C_{in}} \qquad (9.17)$$

of the gate is a measure of the loading placed on it when it is used in a cascade, and is called the **electrical effort**. The time delay, t_d, through the gate is given by

$$t_d = (gh + p)\tau \tag{9.18}$$

where p is due to internal FET capacitances and τ is a reference time constant. All of these parameters depend on the FET aspect ratios. Logical effort provides a method to minimize the delay of a cascade by adjusting the sizes of gates in a chain. Since it is a rather involved technique, the interested reader is directed to the references for the details. [9.4]

9.3   Library Entries

Although there is no standard set of library cells, digital-design requirements dictate a fairly large group of specific functions that a chip designer needs to complete a system design. Some of the more obvious components are listed here, but the list is far from complete.

Primitive Gates

This includes NOT, NAND, and NOR, in addition to AND, OR, and buffers. XOR and XNOR gates are often included in the classification of primitive gates, as are individual transistors of various sizes and transmission gates.

General Logic Elements

Complex components. such as multiplexors, decoders, encoders, comparators, and equality detectors, are useful in digital design.

Latches and Flip-Flops

State elements are required for synchronous design. A typical base set includes D-latches and FFs, SR latches, and different register designs. Both level-sensitive and edge-triggered components are important. Clocking circuits are also included in this group.

Memory Elements

Static RAM (SRAM) cells and arrays, along with associated circuitry, are used extensively. Dynamic RAMs (DRAMs) are more specialized and can be difficult to incorporate.

Arithmetic Components

Bit adders and subtractors, including related circuits such as counters, are used for many applications. Word adders and subtractors are usually included, as are integer multipliers. Floating-point circuits are more specialized and are only found in highly developed libraries.

Logic Arrays

A logic array is a generic logic circuit that can be user-configured to create the desired functions. These range from simple AOI or OAI arrays to full-blown network synthesizers.

LSI Level Components

A library cell can be as complex as needed. Some libraries include word-level ALU's (arithmetic logic units) or even full microprocessor cores. Although these are obviously more complicated to deal with, to the physical designer they are just another (large) cell with (many) input and output ports.

CMOS Specific Elements

Many library entries are circuits that are needed to make CMOS work to its full potential or to interface the circuitry with the outside world. I/O pad designs are almost always library elements, as they can be difficult to design. Line drivers, skew-control networks, arbitration/interface units, and delay elements are often included in the toolset.

Analog Circuits

Increased integration density has pushed many analog functions into the realm of a digital chip designer. Analog-to-digital (A/D) and digital-to-analog (D/A) converters, oscillators, phase- and delay-locked loops (PLLs and DLLs), sense amplifiers, and a host of other cells have worked their way into digital-design libraries. This trend is growing quite rapidly. Many high-speed digital CMOS circuits are based on small-signal differential signals, so that (in the author's opinion) the artificial line between digital and analog CMOS will disappear in the near future.

It is obvious that the contents of a cell library can be quite extensive. Regardless of the number of entries, however, it should be remembered that the concepts are unchanged. Standard cell design uses the cells as building blocks, combining and growing as needed for the problem at hand. Physical-design principles are the same, whether the library has 20 or 1,000 entries. Only the design solutions are different.

● ● ● ● ● ● ● ● ● ● ● ● ● ⋅ ⋅ ⋅ ⋅

9.4 Cell Shapes and Floor Planning

Floor planning is the art of arranging the circuits to create the finished chip. This is usually done using a block viewpoint as portrayed in Figure 9.22. Large-scale functions are grouped together to form blocks that can be pieced together. Interconnect considerations are also important here, since the various sections must be able to communicate efficiently. The size and shape of any logic block depends on the method used to place the primitive entries. In the drawing, the individual registers are shown in the Register file block to illustrate the point. Planning the final shape of a block is an exercise in hierarchical geometry. When cells are assembled from primitives, the size and placement of each entity affects on the final structure. Although this statement is obvious, it can be difficult to create the desired block on the first attempt.

Figure 9.22:
Floorplan example

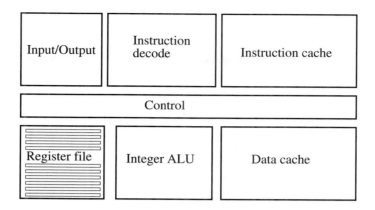

Microwind allows you to construct large circuits. The maximum number of rectangles is set at 20,000, which implies that the layouts can be quite complex. As with any tool of this type, complexity is best handled in a structured manner.

Let us examine the problem using an example. Suppose that we want to create a 2:4 decoder with the logic shown in Figure 9.23. The basic component is the NAND2-NOT or AND gate, as it is used four times in the design. The AND2 cell in Figure 9.24 was constructed using the **Insert Layout** command on the NAND2 gate (*nand2.MSK*) and NOT circuit (*inv.MSK*) in the Microwind folder and then moving and wiring them together. The simplest approach to constructing the 2:4 circuit is to tile four AND2 gates horizontally. This gives the overall layout shown in Figure 9.25. In this case, the 2:4 layout block is short vertically, but long in the horizontal direction.

Figure 9.23:
2:4 active high
decoder

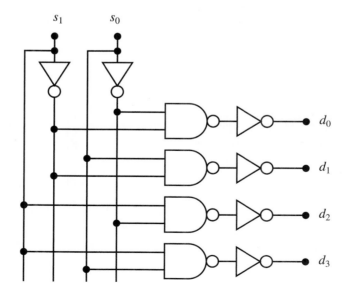

Figure 9.24:
AND2 cell created
from primitives

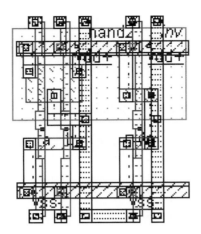

Another choice is to create a 2×2 array, as shown in Figure 9.26. Allowing a wiring channel between the rows gives room for commonly selected signal lines to be inserted between them. This simplifies the wiring to each cell. The outputs are shown at the top and bottom, but they can be routed into the wiring channel if desired. This choice results in a block that is more square; adding the select inverters (one to each row) helps square out the overall shape.

The third choice would be a vertical stack of four cells, as implied by the original logic diagram in Figure 9.23. With the choice of a primitive AND2 cell, however, this would complicate the selected wiring, as the layout would have to follow the diagram very closely.

This simple example illustrates how the placement of primitives in standard cell design affects the size and shape of every block. As units are embedded in more complex structures in the hierarchical-design process, shaping the finished block becomes more and

Figure 9.25:
2:4 decoder with
horizontal layout

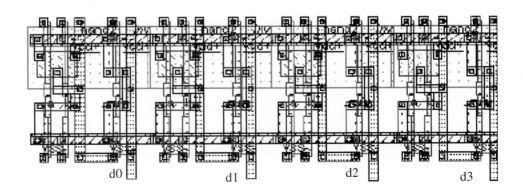

Figure 9.26:
A 2:4 decoder using
a 2 × 2 cell array

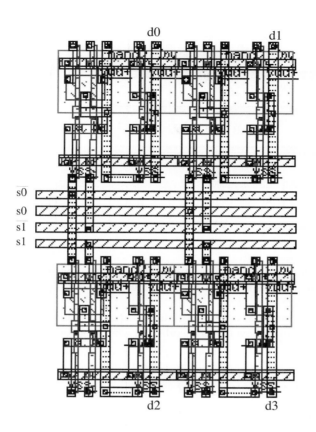

Figure 9.26: A 2:4 decoder using a 2 × 2 cell array

more complex. This is compounded by the routing problem where each section must interface as dictated by the circuit.

Manual planning is often done using simple, proportionately sized blocks. The example in Figure 9.27 shows how various cells can be tiled on horizontal chains and that additional rows add vertical size to the unit.

In Microwind, successive applications of the **Compile One Line** command tends to build horizontal rows that are interfaced directly to VDD and VSS lines. This creates long chains of logic cells, and the order of entry is important, since it is difficult to separate the individual entries. For example, repeating the expression

$$d = A \ \& \ B \tag{9.19}$$

four times automatically gives the layout in Figure 9.25. However, to create a 2 × 2 array with the compiler, first create a 2-cell array with two successive commands, copy it, and then place the copy above the first. Better placement control is obtained using the Insert Layout command, which copies a cell from the library into the present layout area.

Figure 9.27:
Cell placement
planning

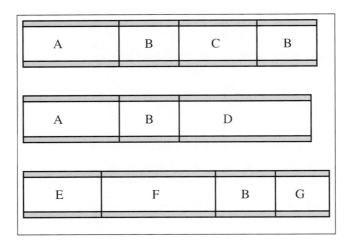

Although the inserted cells are still in a horizontal line, they are separate and can be moved easily and positioned as needed to create the overall pattern. The All button always gives a complete view of the layout.

Every CAD tool has certain procedures that allow the user to create the desired pattern. The trick is learning how the tool works and what its natural tendencies are.

9.5 References

[9.1] Clein, D., *CMOS IC Layout*. Woburn, MA: Newnes, 2000.

[9.2] Martin, K., *Digital Integrated Circuit Design*. New York: Oxford University Press, 2000.

[9.3] Smith, M. J. S., *Application-Specific Integrated Circuits*. Reading, MA: Addison-Wesley, 1997.

[9.4] Sutherland, I., Sproul, B., and Harris, D. *Logical Effort*, San Francisco: Morgan-Kauffman, 1999.

[9.5] Uyemura, J. P., *A First Course in Digital Systems Design*. Pacific Grove, CA: Brooks-Cole, 2000.

[9.6] Uyemura, J. P., *Introduction to VLSI Circuits and Systems*. New York: John Wiley & Sons,. 2002.

[9.7] Wolf, W., *Modern VLSI Design, Third Edition*. Prentice Hall, Upper Saddle River, NJ, 2002.

9.6 Exercises

9.1   Use the Compile One Line function to create a CMOS cell that implements the AO31 operation,

$$f = a \cdot b \cdot c + d$$

9.2   Use the Compile One Line function to create a CMOS cell for the logic function,

$$F = \overline{a \cdot (b + c) \cdot (d + e)}$$

9.3   A useful building block is the Word Equal function. A 4-bit unit accepts two input words,

$$a = a_3 a_2 a_1 a_0$$
$$b = b_3 b_2 b_1 b_0$$

and produces an output of $Eq = 1$ only if $a_i = b_i$ for all $i = 0, 1, 2, 3$; otherwise, $Eq = 0$. Use the Compile One Line command in Microwind to create the logic for the i-th bit, then construct a 4-bit unit.

9.4   Use the Compile One Line function to create a CMOS cell that implements the function

$$f = \overline{a \cdot b \cdot (c + d)}$$

Examine the layout and verify the logic.

9.5   Using the Compile One Line command, create a basic cell library for the following functions:

NOT

AND2, AND3

OR2, OR3

XOR2

Save each file with a unique name, such as *My_not* or *Lib_NOT*. Be sure to not use a name that is already in the Microwind folder. Then use the Insert Layout command to construct layouts for the following functions on a cell-by-cell basis.

(a) $g = (a \cdot b + c) \cdot \overline{(a \oplus b)}$

(b) $f = (x + y) \oplus (a \cdot b)$

(c) $h = \overline{a \cdot b \cdot c + (u \oplus v) + x}$

Try other functions that you may think of for extra practice!

9.6   A 2:1 MUX is described by the output:

$$Out = P_0 \cdot \bar{s} + P_1 \cdot s$$

(a)   Construct the logic diagram using AND and OR gates.

(b)   Use the Compile function to create CMOS cells for AND and OR functions, then use them as primitives in building the MUX.

9.7   An active low decoder produces one line that is low while the others are high. Design a 2:4 active low decoder by starting with the logic diagram and then creating a horizontal layout with the Compile One Line command.

9.8   Design a 3:8 active high decoder using the following procedure.

(a)   Define the select word as $s_2 s_1 s_0$. Then write each output in terms of the select word, e.g., $d_0 = \bar{s}_2 \cdot \bar{s}_1 \cdot \bar{s}_0$.

(b)   Design a primitive AND3 CMOS cell.

(c)   Use eight AND3 cells to design the decoder unit. Remember to consider the wiring and NOT operation needed for the select bits.

CHAPTER 10

Storage Elements—Design and Layout

Modern digital design relies on the ability to monitor and hold the values of binary variables. This chapter examines some common CMOS storage circuits, including basic latches and flip flops. While the treatment is not exhaustive, it is designed to illustrate the physical design of some common elements.

10.1 SR Latch

The SR latch has two inputs, S and R, that set or reset the value of the output, Q. We usually include the **complement**, $\overline{Q}$, as an output to provide maximum flexibility in its usage.

Consider the operational description shown in Figure 10.1. If $(S,R) = (0,0)$, then the latch is in a *hold* state, which means the Q and $\overline{Q}$ retain their current value. If (S,R) changes to $(1,0)$, the inputs set the latch outputs to $(Q,\overline{Q}) = (1,0)$; returning to $(S,R) = (0,0)$ holds this value. Conversely, if (S,R) changes to $(0,1)$, the outputs are reset to $(Q,\overline{Q}) = (0,1)$, which is held when the signal returns to $(S,R) = (0,0)$. The combination $(S,R) = (1,1)$ is not used, as it causes a conflict in the circuit definitions of Q and $\overline{Q}$. Q and $\overline{Q}$ cannot both be the same logic state of "0".

Figure 10.1: Operation of an SR latch

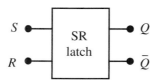

S	R	Q	$\overline{Q}$	
0	0	Q	$\overline{Q}$	→ Hold
1	0	1	0	→ Set
0	1	0	1	→ Reset
1	1	0	0	→ Not Used

Figure 10.2:
NOR-based latches

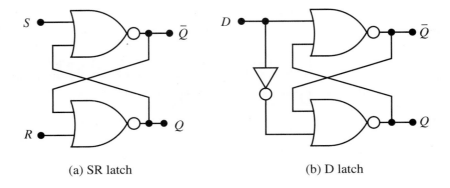

(a) SR latch   (b) D latch

A NOR implementation of the SR latch is shown in Figure 10.2(a). Adding an inverter, as in Figure 10.2(b), yields a D-type latch. The output Q of a D latch follows the input D after a switching delay. Both circuits are based on a pair of cross-coupled NOR2 gates. The layout in Figure 10.3 uses common V_{DD} and V_{SS} connections in the center and gives a NOR gate on either side. The cross-coupling is achieved using horizontal metal lines in the center regions between the two NOR transistors. We have left the inputs and outputs on vertical Metal2 lines at this stage.

The layout for a D latch is shown in Figure 10.4. This design was obtained by adding an inverter to the right side of the SR latch, then wiring the NOT output to an input poly on the right NOR2 circuit. The inputs have been routed along the top using horizontal Metal3 lines, while the outputs are taken from the bottom of the circuit.

Figure 10.3:
SR latch using
NOR2 primitives

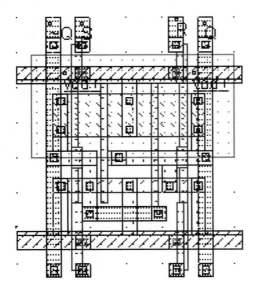

Figure 10.4:
D-latch layout from
the SR-latch circuit

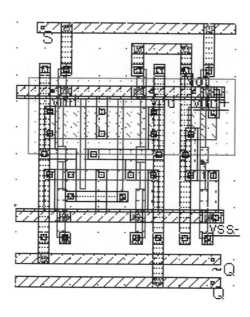

A clocked SR latch is shown in Figure 10.5. This uses a clock input ϕ that controls the AND gates. When $\phi = 1$, the latch is enabled, since the (S,R) inputs can be transmitted through to the cross-coupled latching circuit. If $\phi = 0$, the inputs to the NOR2 gates are automatically 0, which places the latch in a hold state. Although we can add AND gates to the basic SR circuit, the partitioning shows that we only need to wire two AOI21 gates. This gives the layout shown in Figure 10.6. This was created by starting with the **Compile** feature for the function

$$AOI = \sim A|(B\&C) \tag{10.1}$$

Figure 10.5:
Clocked SR latch

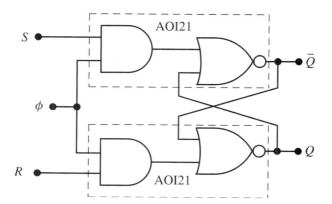

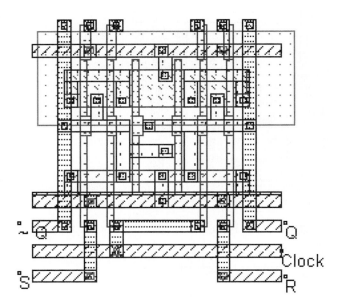

to create a basic gate. A copy was made and flipped horizontally, then merged together into a single cell. Superfluous contacts were eliminated before combining the two cells. To achieve the cross-coupled wiring, the nFETs were moved downward (closer to V_{SS}), so that Poly Metal1 contacts could be added in the center of the cell. Metal3 lines were used in the lower routing channel to complete the design.

● ● ● ● ● ● ● ● ● ● ● ● ● ● ● ●

10.2 Bit-Level Register

A simple register can be built using two inverters and two transmission gates. Figure 10.7 shows the basic circuit. The transmission gates are clock controlled. When $\phi = 1$, TG1 is closed while TG2 is open. This allows the input bit, D, to enter the circuit such that

$$Q(t + t_d) = D(t) \tag{10.2}$$

where t_d is the delay through the path. When the clock switches to $\phi = 0$, TG1 is open so no new values of D can enter. TG2 is closed and allows the two-inverter chain to hold the value. The device is a level-sensitive latch that is transparent when $\phi = 1$. The standard-logic symbol is shown in Figure 10.8.

Figure 10.7:
Transmission gate
D latch

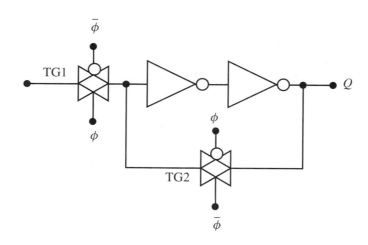

Figure 10.8:
Clocked D latch

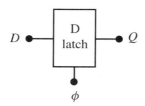

Figure 10.9:
Bistable circuit

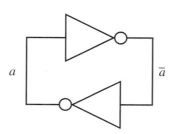

The hold mechanism is due to the properties of the **bistable circuit** that is formed by the two inverters. This is shown in Figure 10.9. The closed loop consists of two inverters, such that both $a = 0$ and $a = 1$ are stable states, i.e., they will hold their value. This is due to the feedback and can be verified by tracing through the loop. Any closed loop that has an *even* number of inverters will be a stable circuit. If an *odd* number of stages is used, the ring may oscillate, and it appropriately is called a **ring oscillator**.

A similar circuit replaces TGs with nFETs and is shown in Figure 10.10. This gives simpler wiring, as both TG pFETs and their clocking signals are eliminated. Since nFETs

Figure 10.10:
nFET-based D latch

Figure 10.10:
nFET-based D latch

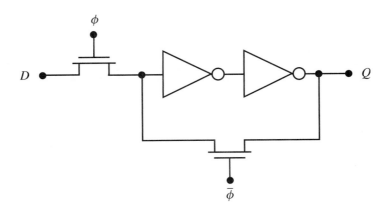

cannot pass a strong logic 1 voltage, the response may be slower, unless large pass transistors are used. Since the bistable circuit is formed during a hold condition, $\phi = 0$, this latch is often the preferred design because of its compact design.

Figure 10.11 is a layout for a D latch. Two inverters share VDD and VSS connections in the center of the cell. The D input is from the left through a transmission gate, while feedback is obtained using an nFET on the right side. The output Q is taken from the right inverter, and $\overline{Q}$ is also available in this design. The latch is **transparent** in that the output follows the input when the clock is high.

Figure 10.11:
Layout for a D latch
with an nFET
feedback transistor

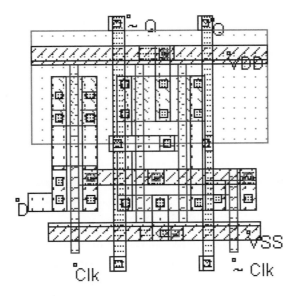

Figure 10.12:
DFF symbol

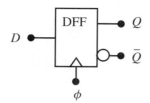

10.3　D-Type Flip-Flop

A flip-flop is an **opaque** storage element; there is no relationship between the output $Q(t)$ and the present value of the input $D(t)$. A **positive-edge triggered D-type flip-flop (DFF)** symbol is shown in Figure 10.12. This loads the value of D on when the clock makes a positive transition from 0 to 1. The edge-triggered property is indicated by the "triangle" at the clock input.

In CMOS, the simplest approach to building a DFF is to use a master–slave arrangement. The input is fed to a "master" D latch and is transferred to a "slave" latch when the clock transition takes place. An nFET-based master–slave flip-flop (FF) is shown in Figure 10.13. Since the input nFET is controlled by $\overline{\phi}$, the master accepts the input when $\phi = 0$. During this time, both the feedback transistor in the master latch and the input FET to the slave are open. When the clock makes a transition to $\phi = 1$, corresponding to a positive edge, the master input is blocked and the bit is transferred to the slave. The master feedback loop is closed to ensure complete transmission. Note that we have rearranged the latch circuit so that the data bit only sees two inverters. This speeds up the latching mechanism.

Figure 10.13:
Master–slave
D-type flip-flop

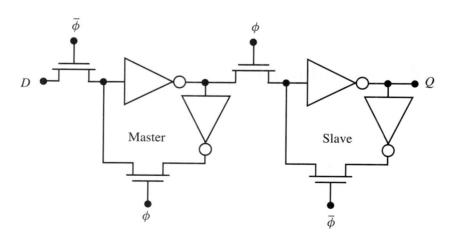

Figure 10.14:
Layout for
DFF circuit

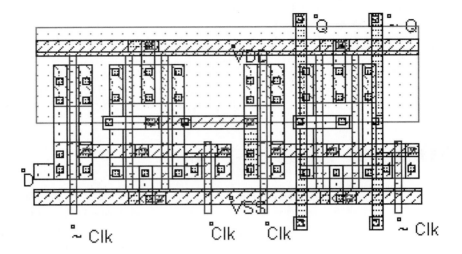

The Microwind layout in Figure 10.14 represents a basic approach to the layout of the DFF. This is based on a modified D-latch cell that was replicated and then wired together. TGs are used for the latch inputs to ensure full rail transmission at a faster speed. The feedback loops use only nFETs, although there is sufficient room to change these into full-transmission gates if desired. Also note that some I/O ports to the internal nodes have been eliminated.

Other flip-flop designs exist and they are important components in a cell library. In classical digital design, FFs provide the basis for important units, such as counters and registers.

10.4 Dynamic DFF

A **dynamic circuit** operates by using the parasitic capacitance on a CMOS node to store electric charge.

A **dynamic flip-flop** can be built using two oppositely phased tri-state inverters, as shown in Figure 10.15. These are connected in a **master–slave configuration**. When $\phi = 1$, the first stage is active, and the voltage V on C is set by the input d. If $d = 0$, then $V = V_{DD}$, while an input of $d = 1$ gives $V = 0$ V. The value is held on the capacitor when the clock changes to $\phi = 0$, since this drives Stage 1 into a Hi-Z state, while simultaneously activating Stage 2. The output, *Out*, is available at this time. Since the data input is latched when the clock makes a transition from $\phi = 1$ to $\phi = 0$, this acts as a **negative-edge trig-**

Figure 10.15:
Dynamic DFF

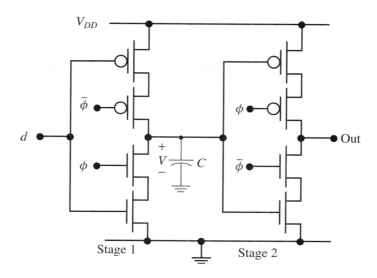

Figure 10.15:
Dynamic DFF

gered (master–slave) **DFF**. The negative-edge trigger characteristic is denoted on the symbol by adding a bubble to the clock input.

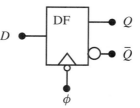

Negative-edge triggered
FF symbol

    While the operation appears to be straightforward, charge storage on the capacitor, C, is subject to a problem called **charge leakage** that exists in CMOS circuits. This is due to unwanted conduction paths, like subthreshold and reverse pn-junction currents, that cannot be eliminated. In practical terms, this means that the clock frequency must be high enough to induce a latching event before the charge state changes. Figure 10.16 shows the layout for this circuit. It was designed using a **tri-state inverter circuit** as a basis for replication. The clocking signals (denoted as "Clk" and "~Clk" in the layout) are obtained from external clocking sources that have been run as horizontal Metal3 lines.

    The dynamic circuit can be modified to a static clocked design by inserting a regular inverter in between the stages and using the second dynamic stage as a feedback control circuit. The schematic is shown in Figure 10.17. In this circuit, the input stage is clocked and passes the bit voltage to the static inverter. The circuit does not suffer from charge-leakage problems because the static inverter provides direct connections to both V_{DD} and ground.

Figure 10.16:
Dynamic FF layout

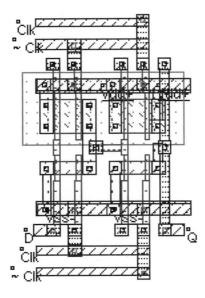

Figure 10.17:
Modified latch for
static operation

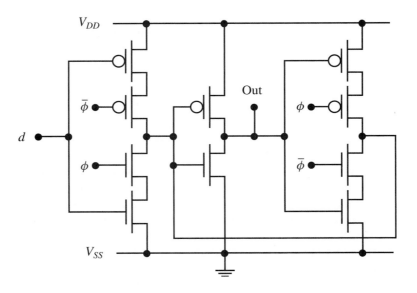

Figure 10.18:
SRAM cell

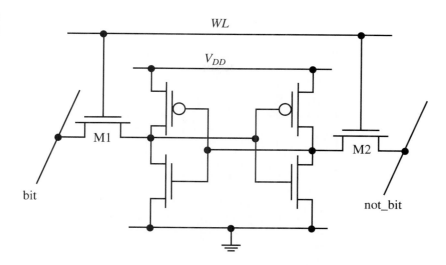

10.5   The Static RAM Cell

Static random-access memories (SRAMs) are highly repetitive VLSI structures that are used for read/write data storage. An SRAM cell is different from a simple latch, in that it uses the same lines for input and output.

The most widely used design is the 6T (six transistor) circuit shown in Figure 10.18. Bit storage is obtained using the bi-stability property of the cross-coupled inverter circuit in the center. M1 and M2 are called **access transistors** and are used to provide conduction paths to the internal bit storage circuit. The access FETs are controlled by the **word line signal**, *WL*. When *WL* = 0, both transistors are OFF, and the cell is in a *hold* state. To perform a read or write operation, the word-line signal is set to *WL* = 1. The SRAM cell uses a **dual-rail** data I/O scheme, where the data bit and its complement are used. To write a data bit, *d,* to the cell, *d* is placed on the bit line and $\overline{d}$ is applied to the not_bit line. These values then establish the internal state of the bistable circuit. A read occurs when the internal states are transmitted out of the cell via the **access transistors**.

An example of an SRAM cell is that found in the Microwind file *Ram1.msk*. This is shown in Figure 10.19. The cross-coupled inverters can be identified in the center part of the cell. The **bit** and **not_bit** lines are denoted as **Data** and **n_data** and are vertical Metal3 lines in this design. The word line (labeled as **Sel** in the layout) runs horizontally at the bottom of the cell using Metal1. A simulation of the cell yields the plots shown in Figure 10.20. Note how the **Sel** (*WL*) signal controls the write operation: the inputs are only accepted when **Sel** is high. A write 1 operation takes place in the first 6 ns of operation, as seen by the **Data** line pulses, one of which occurs when **Sel** is high. A read operation occurs during the time interval from about 8 to 15 ns. This transfers a high voltage out to **Data** and a 0-volt level to **nData**. Note the drop on the **Data** side during a logic 1 read.

Figure 10.19:
SRAM layout
example

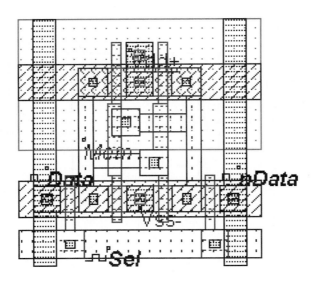

Figure 10.20:
SPICE simulation of
the SRAM cell

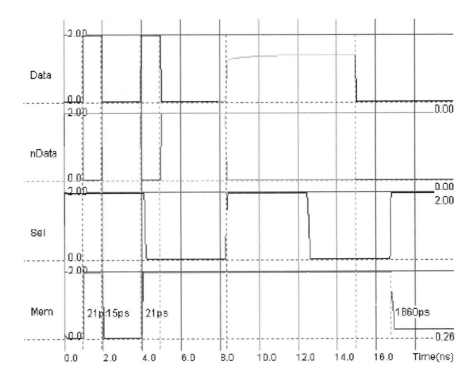

This is due to the threshold voltage drop through the nMOS access transistor. For times greater than 15 ns, the **Data** line is a 0 volt, so a logic 0 is written into the cell when **Sel** goes high at about 17 ns.

10.5.1   SRAM Arrays

RAM arrays are important units in digital design. For example, if we want to store the 4-bit word $d_3d_2d_1d_0$ we need four separate cells. A RAM array consists of many rows, each of which can be broken down into convenient word groups. In CAD tools, the arrays are created by duplicating a single cell. This gives the important concepts of repeatability and regularity.

Microwind has very powerful array generation capabilities that are accessed with the Menu command sequence

Edit ⟹ Duplicate XY

as shown in the screen shot in Figure 10.21. Once this is activated, use the cursor to draw a box around the section that is to be replicated. For our example, we will assume that the SRAM layout is on the screen and that it is enclosed entirely in the drawn box. When the box is defined by releasing the mouse button, the dialog window shown in Figure 10.22 appears. This is used to specify the number of rows and columns of the duplicated objects.

Let us create an SRAM array that has two rows of four individual cells. Inputting $X = 4$ and $Y = 2$ as shown and pushing the **Generate** button yields the array shown in Figure 10.23. Note that the cells are automatically aligned to one another, so that port placement in the original cell is important. The electrical continuity can be checked using the **View Node** operation. The array is easily expanded using other values in the dialog window. A four row, 8-bit storage array is shown in Figure 10.24. This clearly illustrates how CAD tools allow the physical design to implement very complex systems.

Figure 10.21: Accessing the Duplicate XY command

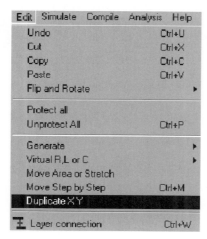

Figure 10.22:
Duplicate XY
dialog window

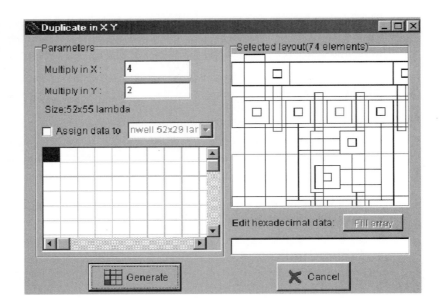

Figure 10.23:
SRAM array
generated by the
Duplicate XY
operation

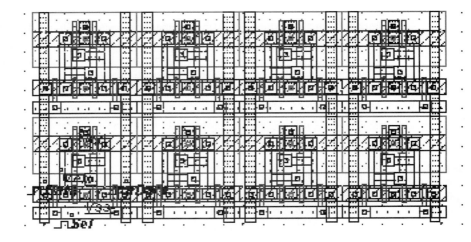

Figure 10.24:
SRAM array for
~~f~~ words

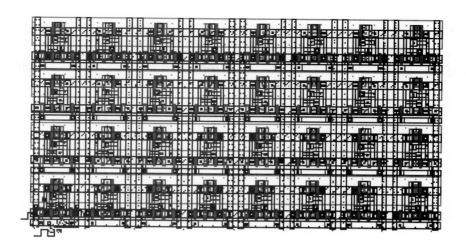

● ● ● ● ● ● ● ● ● ● ● ● ● ● ●

10.6   References

[10.1] Chandrakasan, A., Bowhill, W. J., and Fox, F. (eds.), *Design of High-Performance Microprocessor Circuits*. Piscataway, NJ: IEEE Press, 2001.

[10.2] Itoh, K., *VLSI Memory Chip Design*. Berlin: Springer-Verlag, 2001.

[10.3] Keeth, B., and Baker, R. J., *DRAM Circuit Design*. New York: IEEE Press, 2001.

[10.4] Uyemura, J. P., *CMOS Logic Circuit Design*. Norwell, MA: Kluwer Academic Publishers, 1999.

● ● ● ● ● ● ● ● ● ● ● ● ● ● ●

10.7   Exercises

10.1 Using the NOR implementation of the SR latch as shown in the circuit in the figure below, what are the values of Q and $\overline{Q}$? What are they when $S = R = 1$?

Problem 10.1

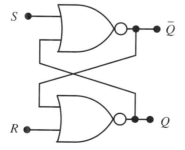

10.2 In the circuit in the figure below, how long can the circuit hold its state if the leakage current from the dynamic node is 10nA? Assume a maximum allowable voltage drop of $V_t = 0.7$ V and $C = 500$ fF.

Problem 10.2

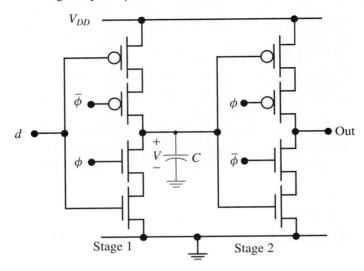

10.3 Using Microwind, simulate a 3-stage ring oscillator in 0.12 μm and 0.25 μm technologies. What is the speed difference?

CHAPTER 11

Dynamic Logic Circuits— Basic Principles

Dynamic CMOS logic circuits provide a different approach to designing high-speed cascades. The philosophy eliminates most of the slow pFETs and employs a clocking signal for both the operation of the gate and data synchronization. Dynamic logic circuits can be very tricky to design because of their complex electrical characteristics. And, in some cases, they can dissipate large amounts of power. This chapter is an introduction to the basics, with an emphasis on the physical design. The interested reader is directed to the reference list for more detailed discussions.

11.1 Basic Dynamic Logic Gates

The general structure of an nMOS dynamic logic gate is shown in Figure 11.1(a). This uses a single clock-controlled complementary pair consisting of Mp and Mn. The logic is performed entirely by an array of nFETs that acts like an open or closed switch, depending on the inputs. This is possible because only one type of FET array (nFET or pFET) is really is needed to provide the switching.

The operation of the circuit is controlled by the **clocking signal** ϕ. When $\phi = 0$, the circuit is in **precharge (P)** where Mp is ON and Mn is OFF. This allows the output capacitor, C_{out}, to charge to a voltage of $V_{\text{out}} = V_{DD}$, every half clock cycle. Logic is performed when the clock transitions to $\phi = 1$, which defines the **evaluate (E)** portion of the operational cycle. During this time, Mp is OFF while Mn is ON, and the inputs a, b, and c are accepted into the nFET logic array. If the array acts like an *open switch*, then V_{out} is held at V_{DD} corresponding to a logic result of $f = 1$. The alternate situation is when the nFET array acts like a *closed switch* and allows the output capacitor, C_{out}, to go to a voltage of $V_{\text{out}} = 0$ V; this corresponds to an output of $f = 0$. As shown in Figure 11.1(b), this sequence occurs with every clock cycle.

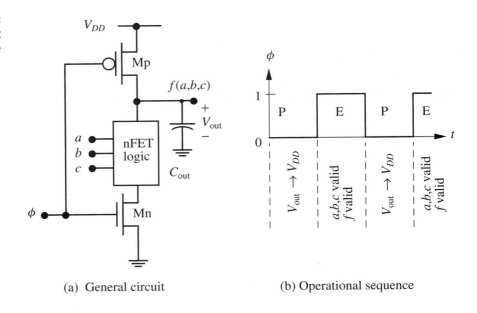

(a) General circuit   (b) Operational sequence

An example of a dynamic logic gate is the NAND2 gate shown in Figure 11.2(a). The logic array consist of two series-connected nFETs with inputs a and b. During precharge, ($\phi = 0$), C_{out} charges to a voltage of $V_{out} = V_{DD}$. When the clock changes to $\phi = 1$, the circuit goes into evaluation. If either input (or both) is 0, then the FET array is an open switch and the voltage is held at the output corresponding to $f = 1$. However, if $a = b = 1$, then both logic nFETs and the clocking transistor, Mn, are ON, which allows C_{out} to discharge to ground. This results in an output of $V_{out} = 0$ V, corresponding to $f = 0$. The circuit is said to undergo a **conditional-discharge** event during evaluation.

The circuit layout is straightforward and is shown in Figure 11.2(b). Three nFETs are connected in series with the one closest to ground (on the left) used as the clocking device. A common gate is used for the clocked complementary pair, but only single nFET inputs are needed for the logic array. This shows one advantage of dynamic logic gates: simpler wiring. Eliminating the complementary pFET devices makes the physical design much simpler.

Figure 11.3 shows a SPICE simulation of the NAND2 circuit. The clocking signal (top plot) starts at 0, which charges the output node (bottom plot) to a value of $V_{DD} = 1.2$ V. The clock makes a transition to a 1 at time $t = 1$ ns; the inputs (denoted as pulses in the plot) are then simultaneously pulsed to logic 1 voltages. This causes the output voltage to fall corresponding to $f = 0$. This completes the first logic cycle. The clock returns to a precharge (0) state at $t = 2$ ns, which again causes the output voltage to rise to 1.2 V. During the next evaluate event (3 to 4 ns), the inputs are both 0 and the output is held high to give $f = 1$. This sequence illustrates how the clock synchronizes the operation of the gate with the data input/output timing.

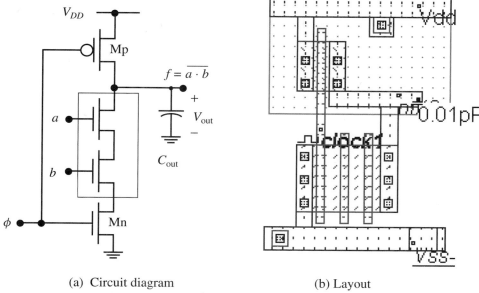

(a) Circuit diagram   (b) Layout

Figure 11.2: Dynamic NAND2 gate

The design technique can be extended to other logic functions using the same nFET arrays that were introduced for static logic circuits. A NOR2 gate is shown in Figure 11.4(a); this produces an output of

$$f = \overline{a + b} \tag{11.1}$$

that is valid only during the evaluation $\phi = 1$. An AOI22 gate that implements

$$h = \overline{a \cdot b + c \cdot d} \tag{11.2}$$

is drawn in Figure 11.4(b). The variations in gate layout revolve around wiring the logic array, since the clocked complementary pair are found in every circuit. It easily is seen that any logic function that can be built in standard CMOS can be implemented in dynamic logic. Dynamic logic circuits are less cumbersome to wire due to the absence of the pFET logic array. They can be fast, but dynamic circuits may consume more power than an equivalent static design.

The basic dynamic logic gate discussed here seems to work fine, but subtle problems in charge storage and transferal are encountered when the logic function gets complex or the gates are used to build a dynamic logic cascade. Various solutions have been developed over the years, but only two are discussed here.

Figure 11.3:
Simulation of the
NAND2 gate

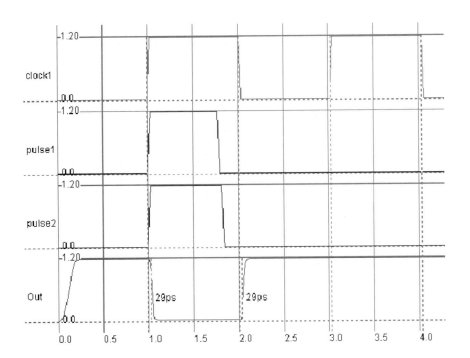

Figure 11.4:
Dynamic logic gate
examples

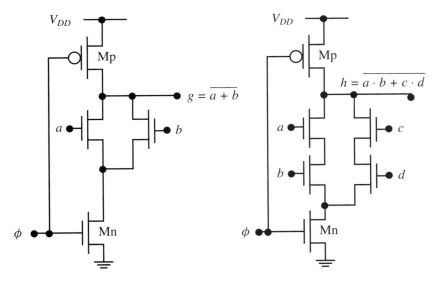

(a) NOR2 circuit

(b) AOI22 logic gate

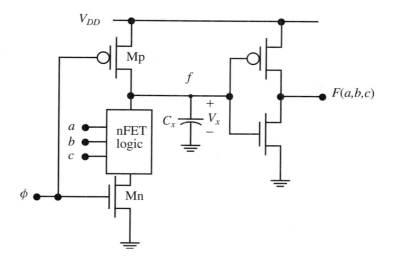

Figure 11.5:
Structure of a
domino logic
stage

● ● ● ● ● ● ● ● ● ● ● ● ● ● ● ● ●

11.2   Domino Logic

Most modern dynamic CMOS logic circuits have evolved from the basic gate design to solve the charge problems. **Domino logic** is an extension that adds an inverter at the output to overcome the possibility of a hardware glitch.

The general structure of a domino-logic stage is shown in Figure 11.5. The inverter creates an output:

$$F(a, b, c) = \overline{f} \tag{11.3}$$

Since CMOS logic is inherently inverting, the addition of the NOT gate makes domino logic **non-inverting**. This means that it is not possible to obtain an inversion without removing the inverter, but this in turn causes charge problems. While this is a limiting factor in using domino gates, the technique is still used for logic networks where this constraint can be satisfied.

The operation of a domino circuit follows the simpler dynamic gate. A clock value of $\phi = 0$ initiates the precharge event. In this circuit, C_x precharges to a value of $V_x = V_{DD}$, which gives an output voltage of $V_{out} = 0$ V. Logically, $F = 0$ is maintained as long as C_x remains charged. During evaluation ($\phi = 0$), a conditional discharge event, governs the behavior of the gate. If the nFET array is an open switch, V_x remains high (for a short period) and $F = 0$. Conversely, a closed switch nFET array discharges C_x to a voltage of $V_x = 0$ V, which changes the output to $V_{out} = V_{DD}$. This corresponds to a logical output of $F = 1$. A 2-input domino AND gate circuit is shown in Figure 11.6(a). The corresponding layout in Figure 11.6(b) is based on the example shown previously for the dynamic NAND2 gate.

Figure 11.6:
Domino AND2 gate

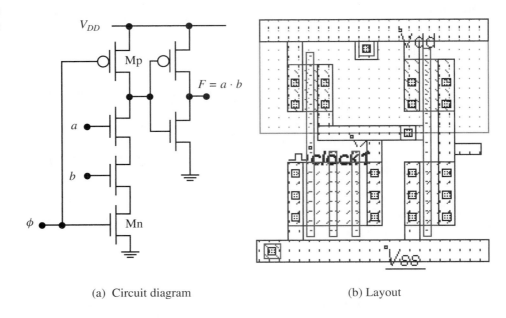

(a) Circuit diagram

(b) Layout

Dynamic logic gates are always used in **cascades**. A domino cascade is shown in Figure 11.7. The output of each stage is connected to a logic FET in the next stage. The inputs to Stage 1 are arbitrary, but the inputs to Stages 2 and 3 are assumed to be from other domino circuits aligned with Stage 1. Note that every stage is controlled by the same clocking signal, ϕ.

Figure 11.7:
Domino cascade

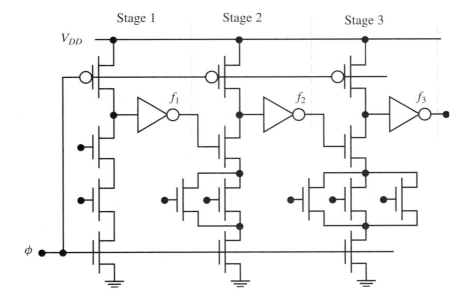

Figure 11.8:
Simulation for a
three-stage domino
cascade

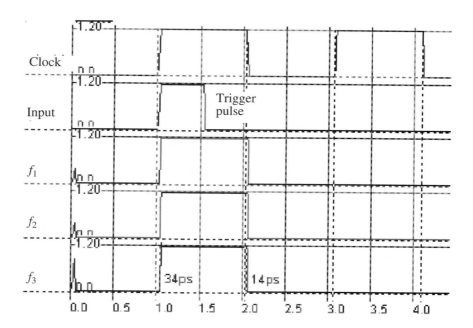

The operation of the cascade is based on the behavior of a single stage. During the pre-charge, $\phi = 0$, and every stage undergoes precharge at the same time. This means that the outputs f_1, f_2, and f_3 are all driven to 0 V, but the outputs are not valid logic outputs. When the clock makes a transition to $\phi = 1$, it starts the evaluation phase by setting up Stage 1 for a conditional discharge. If a discharge does occur in Stage 1, then $f_1 \rightarrow 1$, which turns on the logic nFET in Stage 2. This sets up the possibility of Stage 2 undergoing a conditional discharge event. The logic is thus viewed as "rippling" down the chain, analogous to a string of dominos toppling on one another. This is the origin of the name "**domino logic**." Note, however, that if Stage 1 holds the charge so $f_1 = 0$, it automatically stops the rip-pling, since the Stage 2 discharge path is blocked.

A three-stage cascade was created in Microwind by duplicating the single-stage layout and wiring them together. The simulation plots for the cascade are shown in Figure 11.8. The clock signal (the top waveform) controls the entire circuit. The independent inputs were aligned to the single input that was applied to the first stage.

The waveform set show the behavior of the output nodes. From time $t = 0$ to $t = 1$ ns, every stage is precharged so that the outputs are all 0. When the first evaluated pulse is applied at $t = 1$ ns, the input pulse is applied to Stage 1. The outputs change from 0 to 1 values in a staggered manner: first f_1 changes, then f_2, and finally, f_3. This illustrates the domino effect as the response starts at Stage 1, and this is transferred along the chain.

At $t = 2$ ns, the clock returns to 0 for the next precharge. The second evaluation time starts at $t = 3$ ns. However, since no input pulse is applied during this time, the internal capacitor, C_X, of Stage 1 remains charged, and f_1 stays at 0. This ensures that the logic nFET in Stage 2 remains OFF so that the voltage on its internal capacitor is held. Continu-ing down the chain, we see that all of the outputs (f_1, f_2, and f_3) remain at 0 corresponding to their precharge values. In this case, the output of the cascade will remain at a logic 0 state, regardless of the number of stages. This simple example illustrates the characteristics of the logic evaluation down the chain.

Modern domino logic circuits have evolved to provide faster speeds and reduced power-dissipation levels by using advanced design and signal-control techniques.

● ● ● ● ● ● ● ● ● ● ● ● ● ● ● ● ● ● ● ●

11.3 Self-Resetting Logic

The domino circuit is designed to use the clock pulse to synchronize the precharge event. **Self-resetting logic (SR logic)** uses a feedback network to automatically restore the charge on the internal capacitor after a discharge event.

The basic SR circuit is shown in Figure 11.9. This is seen to be a domino logic gate to which another pFET, MR, has been added. MR is the resetting transistor and is controlled by the gate output through a chain of three inverters. A precharge with $\phi = 0$ charges C_x to $V_x = V_{DD}$, which forces the gate output to $F = 0$. MR is OFF, as can be verified by tracing through the inverter chain. When the clock changes to $\phi = 1$, a discharge event will drain the charge OFF of C_x so that $V_x \rightarrow 0$ V. This causes the output to change to $F = 1$, which is fed to the gate of MR through the inverter chain. MR turns ON and recharges (resets) the **interior node voltage** to $V_x = V_{DD}$, which is the origin of the name.

Figure 11.9:
Structure of a self-resetting logic gate

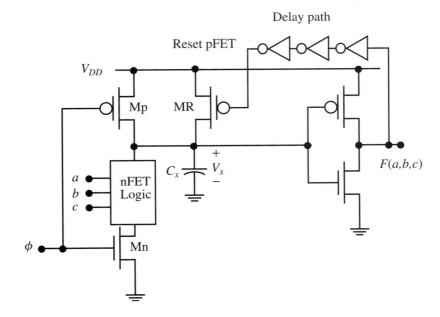

Figure 11.10:
Basic SR gate layout

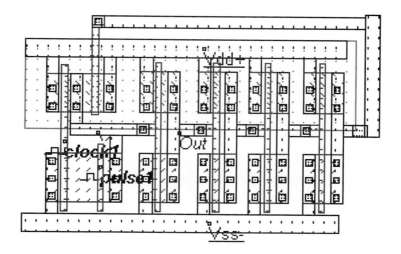

Figure 11.10 shows a simple layout for testing the concept. It consists of a domino circuit with extra inverters and MR added. The simulation plots for the circuit are shown in Figure 11.11. The precharge takes place from $t = 0$ to 1 ns. When the clock changes to $\phi = 1$, the circuit enters the evaluation phase. The input pulse causes the output node to go high, and the internal Node X to fall to $F = 0$ V. During this time, the output voltage is inverted and transmitted back to the gate of the reset transistor. MR turns ON and

Figure 11.11:
Simulation of the self-resetting logic gate

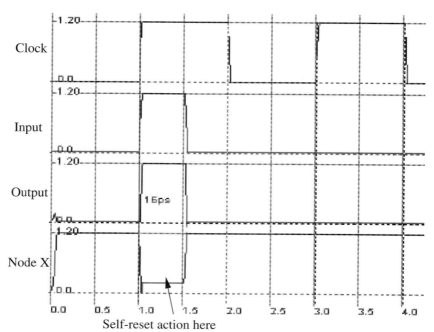

recharges C_x to V_{DD}. Note that resetting Node X back to a logic 1 voltage causes the output F to fall back to a 0 level.

The behavior of this gate is different from others we have studied, in that a logic 1 output is a *pulse*, not a constant level. In this example, the pulse width is less than the evaluation time of the clock itself. During the next precharge event ($\phi = 0$), there is no change because C_x has already been reset. Because the circuit loses its dependence on the clocking, it has been studied for asynchronous applications.

● ● ● ● ● ● ● ● ● ● ● ● ⋯

11.4 Dynamic Memories

Dynamic RAMs (DRAMs) are the most widely used memories because they can be manufactured at the lowest cost-per-bit. System memories, such as those found on the motherboard of your PC, are almost exclusively DRAMs. Modern memory design and manufacturing is based on the best fabrication processes available. Since the size and speed of a DRAM chip relies on the resolution and electrical characteristics of the silicon, they are considered to be fundamental technology drivers.

DRAMs have the lowest cost-per-bit because they are intrinsically small and simple. The basic one-transistor (1T) cell is shown in Figure 11.12. It consists of a single access nFET, MA, and a storage capacitor, C_s. The top of the capacitor is connected to the access transistor, while the bottom terminal is biased with a voltage, V_{bias}. In practice, $V_{bias} = (1/2) V_{DD}$, but we will assume that $V_{bias} = 0$ V for simplicity. When compared with the six-transistor SRAM cell, it is easy to see why DRAM densities are so large.

The operation of the cell is straightforward. To write a bit, the **word line** is elevated to a logic 1 voltage so that $WL = 1$ turns on the MA. The data voltage is applied to the bit line, which sets the charge state on C_s. The **stored charge**, Q_s (in coulombs), is given by

$$Q_s = C_s V_s \tag{11.4}$$

so that logic 0 and 1 values are differentiated by "small" and "large" charge levels. A hold condition is attained by bringing the word line level low to $WL = 0$; this drives MA into cutoff and blocks the main conduction path to the bit line.

Figure 11.12: one-transistor dynamic RAM cell

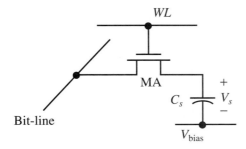

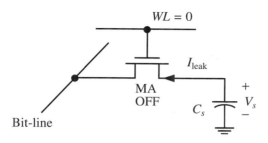

Figure 11.13:
Charge leakage
from a DRAM cell

When a DRAM cell is in a hold state, **leakage currents** through the transistor reduce the amount of stored charge to zero. While this has no effect on a logic 0 (small charge) state, it destroys a logic 1 charge level. The problem is shown in Figure 11.13. The leakage current, I_{leak}, is due to many effects, but the subthreshold conduction is the most important in a submicron technology. Using the *I–V* **relationship** for a capacitor allows us to estimate

$$I_{\text{leak}} = C_s \frac{\Delta V_s}{\Delta t} \tag{11.5}$$

so that the voltage drop ΔV_s is

$$\Delta V_s = \frac{I_{\text{leak}}}{C_s} \Delta t \tag{11.6}$$

where Δt is the time interval that the charge has been held at. Let V_1 be the minimum voltage that can be recognized as a logic 1, and $V_s(0)$ be the initial voltage on the capacitor. The **hold time**,

$$t_h = \frac{C_s}{I_{\text{leak}}}[V_s(0) - V_1] \tag{11.7}$$

is the longest time that a logic 1 can be maintained in the cell; this is also known as the **retention time**. DRAM cells usually have storage capacitors with values around $C_s = 30$ fF. Leakage current levels in submicron transistors vary widely with the process, with a typical value of $I_{\text{leak}} = 1$ nA ($= 10^{-9}$ A) per micron of channel width, W. Assuming $[V_s(0) - V_1] = 1$ V, the hold time is estimated to be

$$t_h \approx \frac{30 \times 10^{-15}}{1 \times 10^{-9}} = 30 \times 10^{-6} \tag{11.8}$$

or 30 μs. The short hold times require us to implement a **refresh operation**, where the contents of the cell are read, amplified, and rewritten at a rate that is sufficient to combat the **charge-leakage reduction**.

Figure 11.14:
DRAM layout
simulation

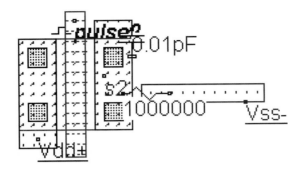

DRAMs can be fabricated only in highly specialized processing lines, because the capacitors themselves are very difficult to create. The large value of C_s cannot be designed in a standard CMOS process without consuming huge amounts of surface area. This would reduce the number of cells per unit area, and the design would lose its high-density advantage. Because of this, it is not possible to design a realistic DRAM cell using the standard technologies in Microwind. We can, however, examine the operation using a virtual capacitor in the layout, as in Figure 11.14. A 0.01 pF = 10 fF capacitor has been added to one end of the nFET to simulate the storage cell. We have also created a 1 MΩ virtual resistor to ground to simulate leakage effects. The signals have been specified to model a Logic1 *write* operation and the charge leakage. Figure 11.15 shows the SPICE results for this circuit. The word line is pulsed, which allows the V_{DD} input to write the logic 1 volt-

Figure 11.15:
DRAM simulation

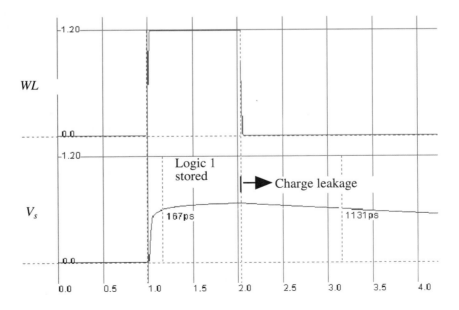

Figure 11.16:
DRAM array
example

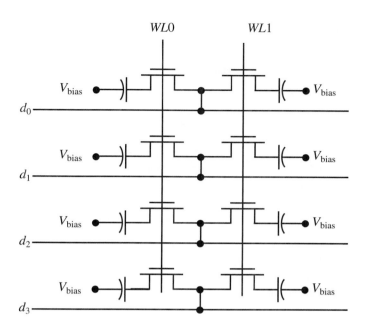

age to the capacitor. The capacitor voltage, V_s, achieves a high level (with the threshold voltage loss quite evident) and then the word line is brought down to $WL = 0$. The leakage then reduces the stored voltage level. It is important that one does not interpret the numerical values in the plots too literally, as this is only a simple, linear model of the problem. Accurate calculation of leakage currents can be achieved only with complex-device modeling. While the BSIM4 models in Microwind are good for most applications, they were simplified for integration into the program and do not include high-level leakage models. For that matter, very few SPICE models before BSIM3 include charge leakage.

DRAM arrays are designed by applying the VLSI concepts of repetition and regularity. An example is shown in Figure 11.16. This uses a 2-bit cell as a basic unit, with horizontal bit lines (d_0, d_1, d_2, d_3) and vertical word lines ($WL0$ and $WL1$).

● ● ● ● ● ● ● ● ● ● ● ● ● ● ● ●

11.5   References

[11.1] Berstein, K., et al, *High-Speed CMOS Design Styles*. Norwell, MA: Kluwer Academic Publishers, 1998.

[11.2] Keeth, B., and Baker, R. J., *DRAM Circuit Design*. New York: IEEE Press, 2001.

[11.3] Uyemura, J. P., *CMOS Logic Circuit Design* Norwell, MA: Kluwer Academic Publishers, 1999.

● ● ● ● ● ● ● ● ● ● ● ● ● ● ● ● ● ⋅ ⋅ ⋅

11.6 Exercises

11.1 Design a dynamic CMOS logic gate for the function:

$$f = \overline{a \cdot (b + c)}$$

Use nFETs that have aspect ratios of at least 6, and a pFET with an aspect ratio of 4. Simulate the circuit and verify its operation.

11.2 Design the layout for a domino logic gate that implements the AND3 operation,

$$g = a \cdot b \cdot c$$

Then cascade the stage into a non-inverting buffer stage and simulate the operation of the chain.

11.3 Construct the circuit and layout for a self-resetting logic gate that implements the AO function:

$$F = a \cdot b + c$$

Simulate the circuit and measure the width of the output pulse. Discuss methods to increase or decrease the pulse width.

Interconnect—Routing and Modeling

Interconnect complexity is often viewed as a limiting factor in VLSI design. Routing of the wires is only part of the problem. Every line introduces electrical time delays that can cause timing problems. Moreover, electrically induced crosstalk may limit the interconnect density, which in turn affects the entire design. In this chapter, we will study these problems from a layout perspective.

● ● ● ● ● ● ● ● ● ● ● ● ● ● ● ● ● ● ●

12.1   Modeling an Isolated Line

Consider the interconnect line shown in Figure 12.1; the material is assumed to be a metal layer, but the analysis also applies to any polysilicon lines that may be used on a chip. The

Figure 12.1:
Interconnect line
geometry

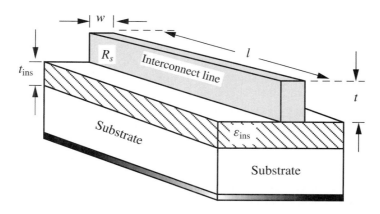

line has a width, w, thickness, t, and length, l, all with units of centimeters (cm). From an electrical viewpoint, the line has three parasitic components: the resistance, R_{line}, the capacitance, C_{line}, and the inductance, L_{line}. At the local level, R_{line} and C_{line} dominate; inductive effects from L_{line} usually are ignored in low-current lines.

12.1.1   Line Resistance

The end-to-end line resistance is given by

$$R_{line} = R_s n \tag{12.1}$$

where R_s is the sheet resistance

$$R_s = \frac{\rho}{t} \tag{12.2}$$

with ρ as the resistivity of the material in units of (Ω-cm), and n is called the number of squares from one end to the other as calculated from

$$n = \frac{l}{w} \tag{12.3}$$

In this definition, a square is defined as having dimensions of $w \times w$ so that it varies with the line width. Because of this, R_s is sometimes said to have units of "ohms per square." The resistance formula can be written in the alternate form

$$R_{line} = rl \tag{12.4}$$

where $r = (R_s/w)$ is the resistance per unit length in Ω/cm.

It is important to note that the sheet resistance is inversely proportional to the thickness of the line, while the resistivity is a fixed parameter for a given material. Table 12.1 gives some typical values for ρ. While aluminum has been used as the primary interconnect material for decades, copper technology has been developed for use in high-speed processes to decrease the line resistance.

Polysilicon exhibits a high sheet resistance even if it is heavily doped. For that reason, poly is not recommended for use as an interconnect. When it is used outside of transistors, some designers will increase the width of the line to decrease the resistance by reducing the number of squares. In advanced processes, the poly is coated with a refractory metal, such as tungsten, to decrease the overall resistivity. This mixture is called a **silicide**, and the poly-coated layer is called a **polycide** or **salicide**. This does bring the sheet resistance down to much lower levels, but it is still higher than a metal. During the process, n+ and p+ regions are also coated, so that their sheet resistances are also decreased.

TABLE 12.1
Resistivity of some common interconnect materials at Room Temperature

Material	Resistivity
Aluminum	27.7 mΩ-μm
Copper	17.2 mΩ-μm
Gold	22.0 mΩ-μm

12.1.2   Line Capacitance

The line capacitance is complicated by fringing electric fields, but can be estimated using the expression:

$$C_{\text{line}} = cl \text{ in Farads (F)} \tag{12.5}$$

where c is the capacitance per unit length. A simple estimate for c that includes fringing effects is

$$c = \varepsilon_{\text{ins}}\left[1.13\left(\frac{w}{t_{\text{ins}}}\right) + 1.44\left(\frac{w}{t_{\text{ins}}}\right)^{0.11} + 1.46\left(\frac{t}{t_{\text{ins}}}\right)^{0.42}\right] \tag{12.6}$$

in F/cm. In this expression, ε_{ins} is the permittivity of the insulator between the interconnect and the substrate such that

$$\varepsilon_{\text{ins}} = \varepsilon_r \varepsilon_o \tag{12.7}$$

with ε_r as the **relative permittivity (dielectric constant)** and $\varepsilon_o = 8.854 \times 10^{-14}$ F/cm as the permittivity of free space. For silicon dioxide, $\varepsilon_r \approx 3.9$, and most SiO_2-based insulating layers have $\varepsilon_r \approx 3.9 - 4.0$.

12.1.3   *RC* Model

The *resistance and capacitance (RC)* of the interconnect line adds an *RC* delay to any pulses that are launched on it. This can be included in circuit simulations using *RC* modeling.

Two basic models are shown in Figure 12.2. The simplest, shown in Figure 12.2(b), uses R_{line} and C_{line} as lumped elements. While this is reasonable for the series resistance, the capacitance is a distributed effect from A to B, so that placing C_{line} at the end overestimates its effect. A better model is shown in Figure 12.2(c). In this approach, C_{line} has been divided into two equal capacitors and one is placed at each end. This can be derived from an analysis of the delay time-constants. [12.3]

Figure 12.2:
Lumped-element
RC models

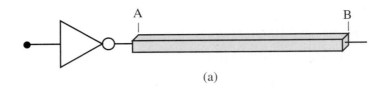

(a)

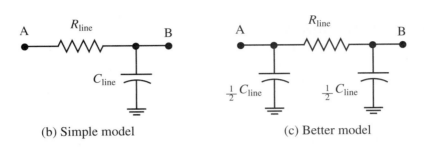

(b) Simple model   (c) Better model

An important characteristic of both models is that they produce *RC* time constants that can be used to estimate signal delays. For the split-capacitor model in Figure 12.2(c), the time constant is

$$\tau = \frac{1}{2} R_{line} C_{line} \qquad (12.8)$$

Using Equations (12.4) and (12.5) expands this to

$$\tau = \frac{1}{2} r c l^2 = \kappa l^2 \qquad (12.9)$$

where κ is a constant. This demonstrates a very important point:

> *The time delay on an RC line is proportional to the square of its length.*

To understand this, first note that a line with length l_1 will have a delay of

$$\tau_1 = \kappa l_1^{\,2} \qquad (12.10)$$

If we design a line that is twice as long with $l_2 = 2l_1$, the delay increases to

$$\tau_2 = \kappa l_2^{\,2} = \kappa (2l_1)^2 = 4\tau_1 \qquad (12.11)$$

i.e., a 4× increase. This illustrates the importance that interconnect layout has on system performance.

Figure 10.6:
Clocked SR-latch
circuit

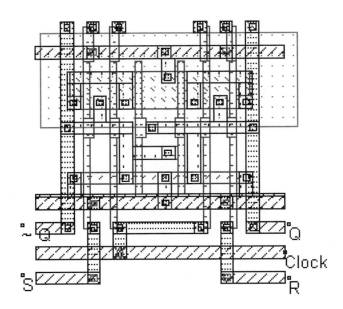

Figure 10.6: Clocked SR-latch circuit

to create a basic gate. A copy was made and flipped horizontally, then merged together into a single cell. Superfluous contacts were eliminated before combining the two cells. To achieve the cross-coupled wiring, the nFETs were moved downward (closer to V_{SS}), so that Poly Metal1 contacts could be added in the center of the cell. Metal3 lines were used in the lower routing channel to complete the design.

● ● ● ● ● ● ● ● ● ● ● ● ● ● · · · ·

10.2 Bit-Level Register

A simple register can be built using two inverters and two transmission gates. Figure 10.7 shows the basic circuit. The transmission gates are clock controlled. When $\phi = 1$, TG1 is closed while TG2 is open. This allows the input bit, D, to enter the circuit such that

$$Q(t + t_d) = D(t) \tag{10.2}$$

where t_d is the delay through the path. When the clock switches to $\phi = 0$, TG1 is open so no new values of D can enter. TG2 is closed and allows the two-inverter chain to hold the value. The device is a level-sensitive latch that is transparent when $\phi = 1$. The standard-logic symbol is shown in Figure 10.8.

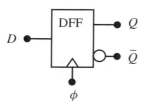

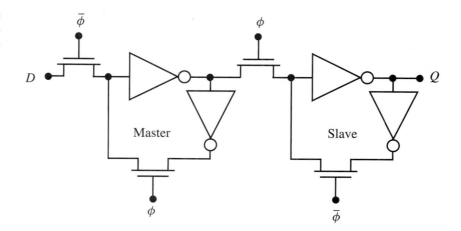

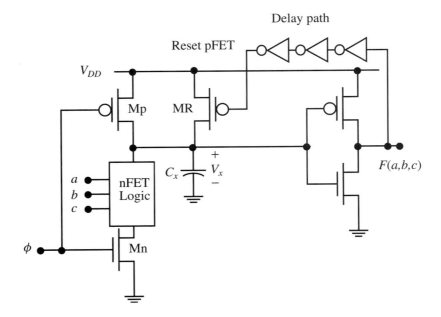

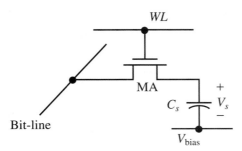

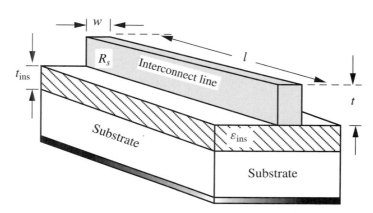

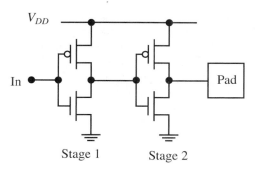

Figure 13.25:
Scaled output-
driver chain

series-connected nFETs and pFETs, so that they must be even larger than the simple circuit to provide the same drive strength and switching times. They are useful for low-frequency outputs where the FETs can be reduced to more acceptable sizes.

The modified tri-state output circuit in Figure 13.27 is controlled by a pair of FETs that make up a **transmission gate (TG) switch**. When $\overline{En}$ = 1, M1 and M2 are OFF and the TG connects M3 and M4 to form the first stage of the output driver. In this case, IN is buffered to the output through the inverter pair Mn and Mp. If $\overline{En}$ = 0, M1 and M2 are ON, and the TG is off. M1 pulls the gate of Mp to V_{DD}, which turns it OFF. Similarly, M2 pulls the gate of Mn to ground, so that both output FETs are in cutoff. This places the output in a Hi-Z state.

The design of high data-rate output drivers is complicated by transmission line effects. As shown in Figure 13.28, a metal trace on a **printed circuit board (PCB)** acts like an LC transmission line with a characteristic impedance

$$Z_o = \sqrt{\frac{L'}{C'}} \tag{13.8}$$

Figure 13.26:
Basic tri-state driver

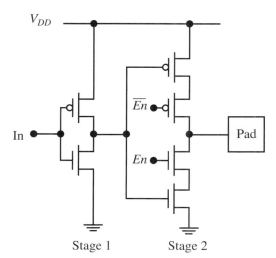

Figure 13.27:
Modified tri-state
output circuit

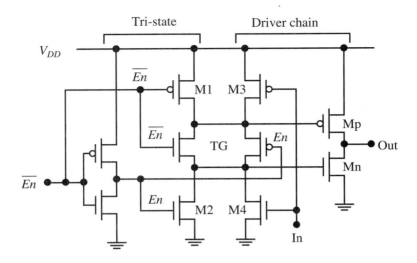

where

> *L' is the inductance per unit length (H/m)*

> *C' is the capacitance per unit length (F/m)*

of the trace. The actual value of Z_o depends on the dimensions of the trace and the dielectric characteristics (thickness, location of ground plane, and permittivity ε). A nominal value is typically $Z_o \approx 30$ to $50 \; \Omega$, but printed circuit boards usually are designed to insure proper electrical continuity, so the values vary.

The matching problem can be understood using the model in Figure 13.29. The output driver is represented by a voltage source, $V_{out}(t)$, in series with the output resistance, Z_{out}.

Figure 13.28:
Origin of
transmission-line
output problem

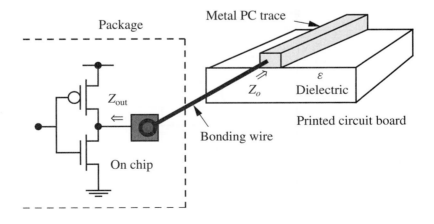

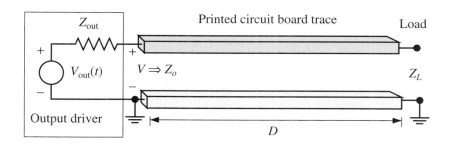

This launches a pulse onto the transmission line created by the PCB trace. Assuming that V_{out} jumps from 0 V to V_{DD} in a steplike manner, the amplitude of the voltage on the line is given by

$$V = \frac{Z_o}{Z_o + Z_{out}} V_{DD} \tag{13.9}$$

since the transmission line looks like an impedance of Z_o. This launches a voltage wave-front that propagates to the load with a phase velocity, v_p. The wavefront reaches the load at time

$$t_d = \frac{L}{v_p} \tag{13.10}$$

The load is assumed to be an open circuit, which is the lowest-order time-domain model for a capacitor. Since Z_L is effectively infinite, the mismatch with Z_o causes a reflection and creates a wave front propagating back to the source. The amplitude of the reflected wave front is also V, and this adds to the voltage V already on the line. The reflected wave front reaches the source at a time, $2t_d$. If $Z_{out} \neq Z_o$, then the source will cause another reflection to take place, which launches another pulse towards the load. Multiple source and load reflections cause the line voltage to change, which is called "ringing" in standard circuit theory. Ringing slows down the transmission, because the load voltage is not well-defined until the line voltage stabilizes.

The driver matching problem usually centers around achieving a matched source with $Z_{out} = Z_o$. If this can be accomplished, then the initial pulse amplitude is

$$V = \frac{Z_o}{Z_o + Z_{out}} V_{DD} = \frac{1}{2} V_{DD} \tag{13.11}$$

With the reflected wave, the line voltage is

$$V + V = V_{DD} \tag{13.12}$$

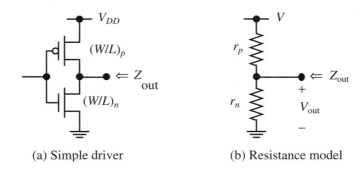

(a) Simple driver   (b) Resistance model

Since there are no secondary reflected waves from the source, this is the final voltage. From the conceptual viewpoint, this is a straightforward problem. The difficulty arises from the fact that the output drivers consist of nonlinear MOSFETs that do not have a single resistance value. This implies that the value of Z_{out} is itself nonlinear, i.e., a function of V_{out}. Simulations easily verify this statement.

Consider a simple inverter, as in Figure 13.30(a). Replacing the FETs with resistors gives the circuit in Figure 13.30(b). However, both r_n and r_p are functions of the voltages, so they vary with V_{out}. This means that it is not possible to match the output impedance to the transmission line with this circuit. One approach is to use a symmetric design with $\beta_n = \beta_p$ and sets $V_{out} = (1/2)V_{DD}$ for the matching point. With the assumption of a transient signal, the power supply acts as an AC ground and the resistors are in parallel. Then we can write

$$r_n = r_p = \frac{Z_o}{2} \tag{13.13}$$

as the matching condition. This allows us to calculate the aspect ratios. Of course, since this is only valid at the midpoint voltage, its behavior will be far from ideal. More complicated approaches add switching FETs to the circuit to vary the impedance as the output voltage changes.

13.3.4   Bi-Directional I/O Circuits

Some pads are used as both input and output ports. A basic circuit for a bidirectional design is shown in Figure 13.31.The input uses the basic ESD protection scheme discussed earlier. The output is a cascade of scaled inverters that are needed to drive the large output capacitance of the pad. The last stage in the chain is a tri-state circuit that is controlled by X. When $X = 0$, the output is in a Hi-Z state and the input stage accepts the signal. The output driver is active when $X = 1$. Of course, any input or output circuit could be used.

Figure 13.31:
Bi-directional I/O
pad circuits

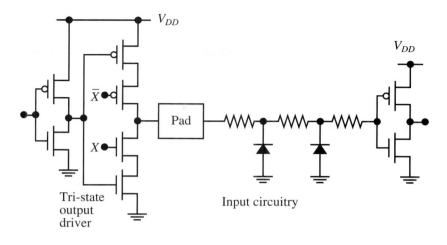

Figure 13.32:
Decoupling
capacitance

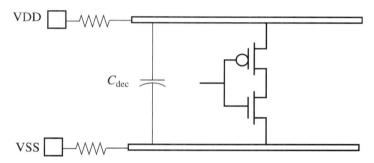

13.3.5   Decoupling Capacitors

Decoupling capacitors are important in VLSI design. As seen in Figure 13.32, C_{dec} is placed between the power supply rails (V_{DD} and V_{SS}). This is used to provide an AC path for any noise signals that exist between the two lines. Random disturbances on the power supply can cause false switching of the logic circuits, so adding C_{dec} usually is mandatory in high density, high speed designs. The actual placement of the capacitor(s) on the chip varies, but the idea remains the same.

● ● ● ● ● ● ● ● ● ● ● ● ● ● ● ● ●

13.4   **The Logo Generator**

Microwind has an interesting feature that allows you easily to create a logo on the chip using a metal layer. The menu command sequence is

Figure 13.33:
Logo Layout
Generator window

Edit ⇒
   Generate ⇒
      Logo

This brings up the dialog window in Figure 13.33. The default letter size is 100 μm, and you can specify the layer where the logo is to be placed. With the input shown, the Logo Layout Generator produces the layout pattern in Metal1 shown in Figure 13.34. This will not be visible in the finished design without etching away the top layers.

It is worth mentioning that the designer should always get permission before putting a logo onto a chip. Also, the large letter size can cause some problems in the planarization process. It is common practice to put fill regions onto a little-used metal layer to make the surface easier to planarize. Since most features are relatively small, the presence of a large logo may affect the fill calculation.

Figure 13.34:
Logo example

● ● ● ● ● ● ● ● ● ● ● ● ●

13.5   References

[13.1] Clein, D.,*CMOS IC Layout*, Boston: Newnes, 2000.

[13.2] Dabral, S., and Maloney, T. J., *Basic ESD and I/O Design*. New York: John Wiley & Sons, 1998.

[13.3] Uyemura, J. P., *Introduction to VLSI Circuits and Systems*. New York: John Wiley & Sons, 2002.

[13.4] Uyemura, J. P., *CMOS Logic Circuit Design*. Norwell, MA: Kluwer Academic Publishers, 1999.

[13.5] Wang, A. Z. H., *On-Chip ESD Protection for Integrated Circuits*. Norwell, MA: Kluwer Academic Publishers, 2002.

● ● ● ● ● ● ● ● ● ● ● ● ●

13.6   Exercises

13.1 Design two CMOS inverters with one using high-speed transistors and the other based on low-leakage devices. Compare the DC and transient characteristics when driving a 0.02 pF load.

13.2 Examine your default Microwind technology for the power supply levels. Then change to a smaller or larger process and check the values of V_{DD} there. Explain why there are differences between the two technologies.

13.3 Layout the circuit in the figure such that the output transistors have aspect ratios of $(W/L)_n = 50$ and $(W/L)_p = 70$, while the first-stage FETs are one-third this size. Comment on the problems encountered.

Problem 13.3

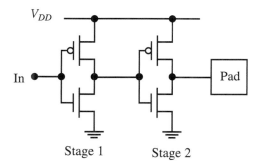

13.4   The breakdown electrical field of a gate oxide is measured to be 8 MV/cm. Find the maximum gate voltage that can be applied to MOSFETs with the following gate-oxide thicknesses.

(a) $t_{ox} = 100$ Å,

(b) $t_{ox} = 50$ Å, and

(c) $t_{ox} = 3.5$ nm

13.5   Generate a 200-Ω resistor in poly, pdiff, and Metal1 layers. Comment on the differences in the resulting structures.

13.6   Design and generate a pad frame with 20 horizontal and 12 vertical pads.

13.7   Use the Logo Layout Generator to create the layout for your name. Compare it to the size of a basic inverter cell.

CHAPTER

14

SOI Technology—
Introduction to Basics

Silicon-on-insulator, or SOI, has been used for various applications, such as radiation hardened circuits, since the 1970s. Recent developments have brought CMOS SOI into the limelight again, this time because of the potential for high-speed design operation. This overview chapter examines some of the main features of this technology.

● ● ● ● ● ● ● ● ● ● ● ● ● ● ●·····

14.1   Modern SOI CMOS

Silicon-on-insulator is a "non-bulk" technology that builds transistors on top of an insulating layer instead of in a semiconductor substrate. This reduces parasitic capacitance levels and yields higher-speed operation, but it introduces a class of problems of its own.

Early SOI technologies used crystals, such as sapphire (**silicon-on-sapphire** or **SOS**), as an insulating substrate. Sapphire was used because its lattice constant (the distance between neighboring atoms) is close to that of silicon, and thin layers of device-grade crystalline silicon can be grown on it. Unfortunately, sapphire and other like crystals are expensive to grow and work with. SOS fell into the realm of specialized circuit families, and attempts to use it for high-speed VLSI designs were never fully realized.

Modern CMOS SOI is based on a more tractable process flow. The objective is still the same: create an insulating layer and build FETs on top. However, SOI starts with a silicon wafer, not a crystal, making it more attractive economically.

An early fabrication sequence for CMOS SOI starts with a wafer that has a thick oxide layer grown on it, as in Figure 14.1(a). An oxide etch Figure 14.1(b) bares the paths to the underlying silicon. A carefully controlled epitaxial growth of silicon creates a "pillar" that has the important crystal structure. This is shown in Figure 14.1(c). The epitaxial process is continued, giving a thin layer of crystal silicon on top of the amorphous silicon-dioxide

Figure 14.1:
Starting sequence
in a simple SOI
CMOS process

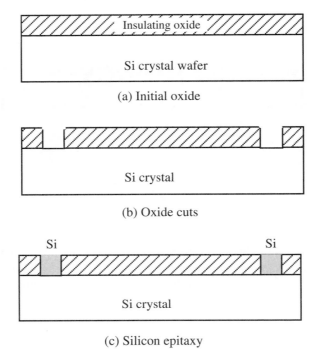

(a) Initial oxide

(b) Oxide cuts

(c) Silicon epitaxy

layer, as portrayed in Figure 14.2. This provides the crystal-silicon background needed for transistors.

A more modern and elegant technique for creating the SOI structure, is to implant heavy doses of oxygen directly into a silicon substrate. This is shown in Figure 14.3(a). The wafer is then annealed at very high temperatures, which induces oxide growth below the wafer surface and "pushes" a top layer of silicon to the top, resulting in the SOI layering shown in Figure 14.3(b). This approach was pioneered by IBM and is called **SIMOX** for "**Separation by IMplantation of OXygen**."

Once the SOI film is made, transistors can be fabricated. Self-aligned MOSFETs are built by defining active-area regions and then ion-implanting dopants are added for polarity and threshold-voltage adjustments. Figure 14.4 illustrates the effect of this sequence. Note that the devices are isolated by etching away the silicon in between active areas and depositing oxide. After planarization, the next steps are to grow the gate oxide, pattern the poly gate, form the spacers, and then implant both the drain and the source dopants. The

Figure 14.2:
Device layer

Silicon layer for transistors

Si crystal

Figure 14.3:
SIMOX process

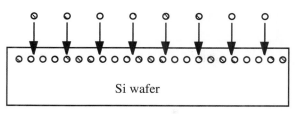

(a) Deep oxygen implant

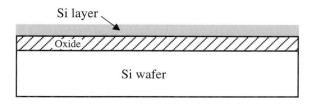

(b) After high-temperature annealing

Figure 14.4:
Active area etching
and background
doping

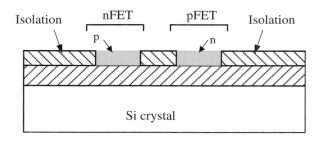

Figure 14.5:
FET formation

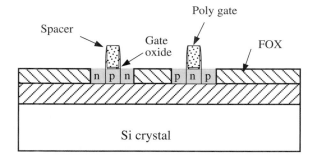

transistors at this point have the cross-section shown in Figure 14.5. A refractory metal coating is used to form the polycide and reduce the contact resistance. These represent the final device structures.

From this point, the process is similar to that in standard bulk CMOS. Oxide is deposited, and contact cuts are formed and then filled to create a plug. The next step is a Metal1 deposition and patterning, which yields the structure illustrated in Figure 14.6. Additional vias and metal patterns are formed in the same manner as in standard processing.

Figure 14.6:
After Metal1
deposition and
patterning

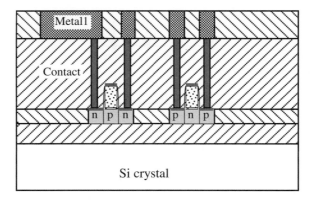

• • • • • • • • • • • • • • • · · ·

14.2 Why SOI?

The basic difference between SOI and a bulk CMOS process is in the placement of the transistors. While silicon epitaxy has been used since the beginning of the semiconductor industry, the technique of creating a single-crystal layer of silicon on top of an amorphous silicon-dioxide layer took many years to develop to the point where it could be moved to the manufacturing line.

CMOS SOI is more expensive than standard bulk processes. As discussed in next section, circuit design is complicated by effects that arise in the physical structure of the devices. The natural question that arises is "What are the advantages of using this technology?" There are two main reasons.

14.2.1 Higher Speed

SOI CMOS exhibits lower capacitance levels than those found in a bulk process. The capacitances of both transistors and interconnects are reduced, which leads to inherently faster switching.

The device bulk capacitances, C_{DB} and C_{SB}, decrease because there is no bottom pn junction formed. A perspective view of an SOI MOSFET is shown in Figure 14.7. Since the entire device is surrounded by oxide, the only pn junctions formed in the structure are beneath the gate. This eliminates the bottom junction entirely, and the only sidewall contribution for a given n+ region is along the channel side. The general expressions for drain–bulk and source–bulk capacitance are then simply proportional to the junction side wall capacitance, C_{jsw}:

$$C_{\text{DB}} = C_{\text{jsw}}W = C_{\text{SB}} \tag{14.1}$$

Figure 14.7:
Junction
capacitance in an
SOI MOSFET

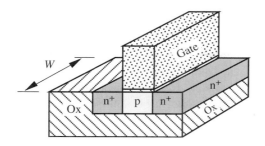

Figure 14.8:
Metal3 capacitance
calculation

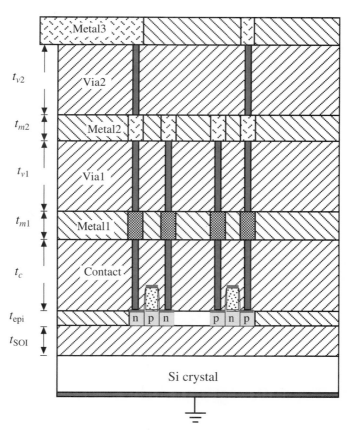

The large C_{bot} contribution has been eliminated. This increases the intrinsic switching speed of the transistors, leading to an intrinsically higher frequency operation. This reduction is significant because it applies to **every** transistor in the circuit.

Interconnect capacitance is also lower when an SOI process is used instead of bulk CMOS, although the reduction is small compared to that found in FETs. To understand this comment, consider the cross-sectional view drawn in Figure 14.8. In this situation,

Metal3 is used as the signal line. The spacing from the bottom of the Metal3 trace to the substrate is given by

$$t_{\text{ins}} = t_{v2} + t_{m2} + t_{v1} + t_{m1} + t_c + t_{\text{epi}} + t_{\text{SOI}n} \tag{14.2}$$

where t_{vx} denotes a via spacing, t_{mx} is an insulator spacing due to a metal layer, t_c is the contact height, t_{epi} is the device layer (epi) thickness, and t_{SOI} is the thickness of the base oxide layer. The interconnect capacitance formula shows us that

$$C_{\text{line}} \sim \frac{1}{t_{\text{ins}}} \tag{14.3}$$

Since the epi and base-oxide layers increase the spacing between the Metal3 line and the grounded substrate, C_{line}, is smaller than in a similar bulk structure. Even though the term $t_{\text{epi}} + t_{\text{SOI}}$ is a small increase in thickness, its effect is enhanced by the reciprocal $1/t_{\text{ins}}$ dependence.

The reduction in line capacitance applies to all conducting layers, with the largest percentage reduction in the lowest interconnect levels. To see this, note that the insulator thickness for a Metal1 line is

$$t_{\text{ins}} = t_c + t_{\text{epi}} + t_{\text{SOI}} \tag{14.4}$$

as compared to bulk CMOS where

$$t_{\text{ins}} = t_c \tag{14.5}$$

Although this line capacitance per unit length will be reduced by a significant amount, Metal1 lines generally are reserved for local wiring and are not used for long interconnects. However, every reduction helps!

These simple arguments illustrate that the overall capacitance levels in an SOI CMOS structure, will be smaller than those fabricated by a standard bulk technology. If a chip design is transferred from a bulk process to an SOI fabrication line, it will run faster. The motivation for accepting the increased expense becomes clear.

14.2.2   Low-Power Operation

Power dissipation in the form of heat is definitely one of the major hurdles for next-generation chip design. A static CMOS logic gate with an output capacitance, C_{out}, dissipates dynamic power according to

$$P = C_{\text{out}} V_{DD}^2 f \tag{14.6}$$

where f is the signal frequency. SOI circuits intrinsically have lower capacitance levels than those built in a bulk technology, so C_{out} will be lower. Extrapolating this argument implies that SOI is a good candidate for low-power designs.

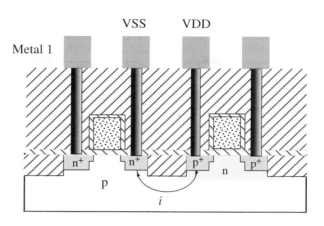

Figure 14.9:
Origin of the latch-up problem

14.2.3   Latch-Up Prevention

Latch-up is an electrical condition that can occur in a bulk CMOS structure. If a chip undergoes latch-up, the current is diverted away from the circuits, and it will fail to operate. Latch-up occurs because a parasitic conduction path is created between the power supply and ground. The path is a consequence of the bulk and well layering used to create both polarities of FETs in a single substrate.

Consider the n-well structure in Figure 14.9. Power supply (VDD) and ground (VSS) connections are shown to illustrate the problem. Tracing the path from VDD to the ground gives the following layering: p+, n-well, p-substrate, n+, and ground. This is equivalent to a 4-layer switching device called a **silicon-controlled rectifier** that is common in power electronics.

Figure 14.10 is a simplified view of the layering. Under normal conditions, the pnpn layering scheme blocks the current flow, and only normal leakage values of i flow. Note,

Figure 14.10:
pnpn layering characteristics

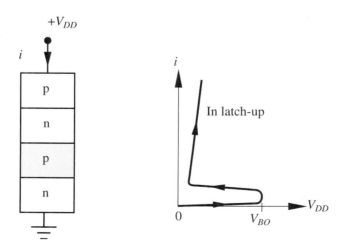

however, the p-region at the bottom. If this region is thin, then a large voltage may swamp this with electrons, making it more like an n-type layer. If that happens, then the structure turns into a forward-biased pn junction, and i will be large. The i–v characteristics show this behavior. If the voltage exceeds the breakover voltage, V_{BO}, then i gets large. If this happens in the CMOS structure, then the chip is in latch-up and the current is diverted away from the transistors to the low-resistance conduction path.

This problem has been studied extensively for many years. In bulk CMOS technology, it is handled by using guard rings, which are heavily-doped n^+ or p^+ rings that surround devices and are biased to block the current. In addition, many layout design rules and circuit techniques have been created to avoid latch-up conditions. The general philosophy is to block the formation of the parasitic conduction path.

SOI CMOS has the advantage that it does not have the bulk connection path between nFETs and pFETs, since it uses the base oxide layer and active area isolation to insulate the transistors. This eliminates latch-up in the structure.

● ● ● ● ● ● ● ● ● ● ● ● ● ● ●
14.3 Problems with SOI

Now that we have seen the advantages of SOI CMOS, let us examine some of the problems. The most obvious one is the increased cost, but this is not viewed as a fundamental limit so long as a fabrication line can produce a profit. The other points are more subtle, and revolve around the structure itself.

14.3.1 Floating Bulks

The isolation and self-aligned MOSFET sequence produces transistors that have hard connections to the drain and source, but there is no inherent connection to the bulk region underneath the gate.

This problem is shown in Figure 14.11. The cross-sectional view in Figure 14.11(a) shows that the bulk forms pn junctions with the drain and the source and is capacitively coupled to the gate through C_G. It is also coupled to the substrate underneath the oxide layer, but this is ignored since the insulator is much thicker than the gate oxide. If we do not bias the bulk, its voltage can vary with the terminal voltages. The problem is complicated by the pn-junction capacitances, C_{SB} and C_{DB}, since they couple time-varying changes to the bulk. It is clear, that letting the bulk voltage float may cause unstable circuit operation.

Floating bulks, give rise to transistors that exhibit hysteresis effects, in that the switching characteristics depend on the voltages set by the previous state. Consider the general threshold voltage

$$V_{Tn} = V_{T0n} + \gamma \left(\sqrt{2|\phi_F| + V_{SB}} - \sqrt{2|\phi_F|} \right) \tag{14.7}$$

Figure 14.11:
Floating bulk

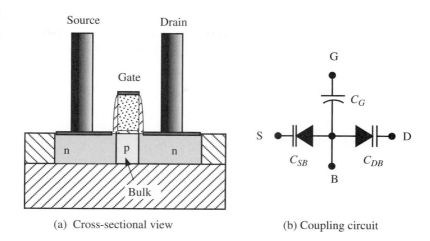

(a) Cross-sectional view   (b) Coupling circuit

Since the bulk voltage, V_B, can vary with the switching, the value of V_{SB} depends on both the present conditions and the previous event. The fastest turn-on will occur with $V_{SB} = 0$. In bulk CMOS, the bulk is grounded so that this only occurs if $V_S = 0$. However, SOI circuits can achieve $V_{SB} = 0$ for nonzero source voltages, since V_B is not pinned to a constant value, accounting for this characteristic in circuit design can be very difficult.

14.3.2   Parasitic Bipolar Transistors

Examining the cross-sectional view in Figure 14.11(a) shows the existence of a parasitic npn bipolar transistor that can affect the device operation. Figure 14.12 shows the parallel-connected devices at the circuit level. The bulk region of the nFET acts as the base of the npn transistor. The collector–emitter voltage is the same as the drain–source voltage: $V_{CE} = V_{DS}$. When the MOSFET conducts current, the bulk can store charge. If the nFET is switched off, the BJT may still continue to conduct collector current using base current

Figure 14.12:
Parasitic BJT

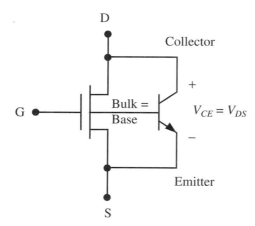

that originates from impact ionization in the p-type region. When combined with the capacitive storage characteristics of the structure, the conditions can make it difficult to turn the current flow off. The pFET also has a parasitic pnp transistor in parallel that causes similar problems.

14.4 SOI in Microwind

Microwind has a sample SOI technology in the file named *soi012.rul* that can be loaded using the **Select Foundry** menu command. This is modeled after a typical process for the purpose of making comparisons.

As an example, consider the simple inverter layout shown in Figure 14.13. This uses small devices and drives an 8 fF (= 0.008 pF) external load. A 2-dimensional cut was made on the nFET with the result shown in Figure 14.14. The main features of the process, including the bulk oxide, device layer, and isolation oxide can be identified.

The transient response was simulated using a clock input and measured at the capacitor. Two Microwind files were used. The first, *cmos012.rul*, is a bulk n-well 1.2-V process that produced the results shown in Figure 14.15(a). The rise time is seen to be 23 ps, and the fall time is 14 ps. The SOI results for the same circuit are shown in Figure 14.15(b); the power supply was reduced from the default 1.5 V to 1.2 V for the comparison. The rise time has increased to 29 ps because the pFETs are optimized for 1.5 V operation, but the fall time has been reduced to 9 ps. Using the **Navigator**, we find that there is only a very small difference of about 0.25 fF in the output node capacitance in the two technologies for this simple circuit. That is because small FETs were used, so there isn't much reduction in junction capacitance.

Figure 14.13:
NOT gate for
simulation
comparisons

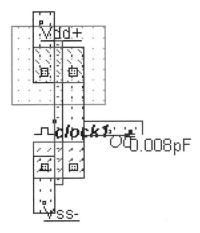

Figure 14.14:
2-dimensional view
of the nFET

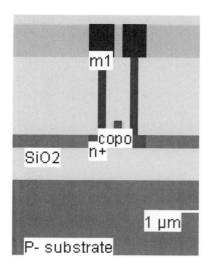

An exercise for the motivated reader is to select a more complex circuit from the Microwind library, and compare the SOI simulation to a bulk process. In general, we expect the difference in speed to increase with the number of transistors and wiring complexity. However, simulations are not always accurate, so some results should be interpreted in a conservative manner.

Figure 14.15:
Inverter simulation
results

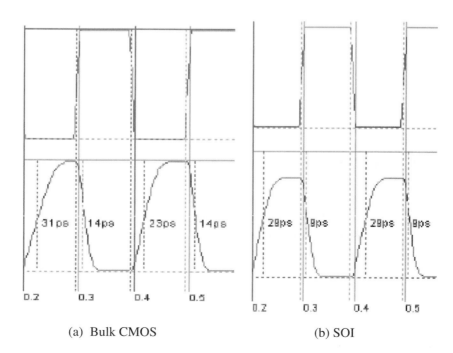

(a) Bulk CMOS   (b) SOI

The Microwind model provides a basic look at modern SOI. Since it is an evolving technology, the details will change as the process becomes more standardized. Updates will be needed to track the new features.

This example serves to illustrate an important point. Progress in creating new silicon technologies is relatively slow. Although CMOS improved dramatically over a 10-year period starting in about 1992, it is getting more and more difficult to continuously design smaller and faster circuits. SOI has been developed for possible use as a future technology base. But there are other competing technologies that are also on the playing field. It usually takes an enormous amount of research and development to take a new process into production.

● ● ● ● ● ● ● ● ● ● ● ● ● ● ● ● ●

14.5   References

Many interesting articles on SOI technology can be found online by performing a search using *Silicon on Insulator* as a keyword group.

The books below provide more detailed discussions.

[14.1] Bernstein, K., and Rohrer, N. J., *SOI Circuit Design Concepts*. Norwell, MA: Kluwer Academic Publishers, 2000.

[14.2] Ecolinge, J. P., et al, *Silicon-On-Insulator Technology: Materials to VLSI*. Norwell, MA: Kluwer Academic Publishers, 1991.

[14.3] Kuo, J. B., and Su, K.-W., *CMOS VLSI Engineering: Silicon-On-Insulator (SOI)*. Norwell, MA: Kluwer Academic Publishers, 1998.

[14.4] Natarajan, S., and Marshall, A. W., *SOI Design: Analog, Memory and Digital Techniques*. Norwell, MA: Kluwer Academic Publishers, 2001.

● ● ● ● ● ● ● ● ● ● ● ● ● ● ● ● ●

14.6   Exercises

14.1   Draw a 4-input NAND gate. Simulate the gate in Microwind using *cmos012.rul* and *soi012.rul*, and compare the results.

14.2   Run the output of the NAND4 from Problem 14.1 through a transmission gate and compare the speeds at the output. How does this compare to the results of the problem? Why? Use device widths of 4 μm for the transmission gate.

CHAPTER

15

Digital System Design 1—
The Dsch Program

This chapter has a different emphasis than those preceding it. Instead of concentrating on chip layout, we will take the discussion to a higher level and study digital system design from logic design down. The process starts with a logic diagram. Using the Dsch program, we can simulate, test, and verify the design and then produce an abstract description of it. The information is transferred to Microwind, which uses the information to produce a CMOS circuit.

The flow discussed here is an example of **design automation**. It revolves around the uses of CAD tools to solve system design problems at all levels. Transferal of the information from one level to the next allows every aspect of the network to be checked and rechecked to ensure that the design is valid.

Our introduction to the topic will be gaining familiarity with the logic design capabilities of Dsch in this chapter.

● ● ● ● ● ● ● ● ● ● ● ● ● ●

15.1   A First Look

Dsch is a companion program to Microwind. It is provided on the CD with this book, and can be loaded onto your PC using the procedure described in Chapter 1.

Launching the program opens the main screen shown in Figure 15.1. The screen dump has been switched to a white background; your screen will have a black background. The layout is very similar to that used in Microwind, with the work area taking up most of the screen. **Menu** commands are listed at the top, and on-screen button shortcuts are also included. The **Palette** menu is replaced by the **Symbol** library. This contains all of the important logic elements that are used to construct schematics.

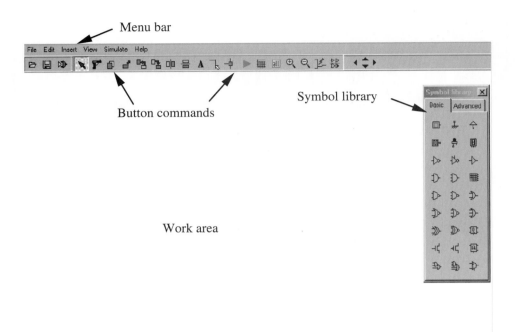

Figure 15.1:
The Dsch launch
screen

The Symbol library has two levels of components that are accessed by clicking on a tab at the top. These are shown in Figure 15.2. The **Basic symbol** set in Figure 15.2(a) provides all of the basic logic elements using shape-specific symbols. The central group of gates are the basic building blocks. This includes (in order):

- NOT, Tri-state NOT, Buffer
- AND2, AND3, Hex keyboard
- NAND2, NAND3, OR2
- NOR2, NOR3, OR3
- XOR2, XNOR2, DFF (Negative-edge triggered)

The next line includes CMOS entities:

- nMOSFET, pMOSFET, D-latch (level sensitive)

The transistors can be used as a basis for drawing CMOS schematics for both digital and analog circuits. The elements in the last line are more general:

- 3-input user defined, 5-input user defined, 2:1 MUX

Figure 15.2:
The Symbol library

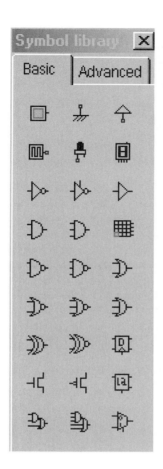

(a) Basic

(b) Advanced

The user-defined characteristic allows one to configure an arbitrary 3- or 5-input gate with a Boolean logic expression. Overall, this collection of symbols is more than sufficient to construct any logic network.

The top buttons of the **Basic Symbol** library are sources and indicators. These are used to simulate and test the circuits, and are detailed in Figure 15.3. The first row provides the following operations:

- Binary input button, Ground (VSS), VDD

The second row is for

- Clock input, LED indicator, 7-segment display

Figure 15.3:
Simulation
elements

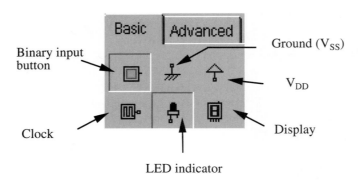

Binary input
button

Ground (V$_{SS}$)

V$_{DD}$

Clock

LED indicator

Display

To test a circuit, you must include inputs (binary input, hex keypad, or a clock) and some indicator output element (LED or display). VDD and VSS are necessary if you are building a logic network using FETs.

The **Advanced Symbol** library is more specialized. It contains components for the inclusion of electrical effects and mixed-signal design. These are used to make a more accurate simulation when the design is transferred to silicon in the form of a CMOS chip.

The components in the Symbol library include many important hardware (chip) features. This is accomplished by linking them to a CMOS process. At the time of this writing, the default is 0.12 μm bulk CMOS, but you should check to see what your version uses. All simulations will exhibit delay times that are intrinsic to the gates designed in the specified technology. The process can be changed using the **Menu** command

Select Foundry

and then choosing the desired technology file.

The **Symbol** library is essential for drawing logic diagrams, but it is not always needed for other tasks. You can close the window using the "**X**" button in the upper-right corner, or it may close itself when certain operations are performed. If it is not visible, you can use the shortcut button shown here to open it. This button is on the extreme right side of the **Menu** bar.

● ● ● ● ● ● ● ● ● ● ● ● ● ● ● ● ●

15.2 Editing Features

The **Edit** menu is shown in Figure 15.4. This provides the same basic functions as those found in Microwind.

Of special note are the 90° rotate buttons. These allow separate clockwise and counterclockwise rotation as these are very useful for orienting the logic symbols when drawing a schematic.

Figure 15.4:
Edit menu

The **Edit** functions are also available on the **Menu** bar. The left group is shown here:

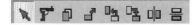

The same icons are found on the **Menu** listing. You will probably discover that editing functions are not needed as often as in Microwind. The drawing process in Dsch centers around using large symbols and *snap-to* wiring. This greatly simplifies the task and eliminates much of the "fine tuning" that is needed in layout design.

● ● ● ● ● ● ● ● ● ● ● ● ● ● ● ●

15.3   Creating a Logic Schematic

At the basic level, Dsch is a schematic entry program. Logic diagrams are created by selecting components from the **Symbol** library, dragging them to the work area, and wiring them together as required. We want to emphasize that Dsch does much more than just provide drawing capabilities, but this is a good place to start.

To construct a schematic, make sure that the program is in drawing mode by pushing the arrow button (shown here) on the left side of the **Menu** bar. A component is placed in the Work Area by the following process:

- Point to the component in the Symbol library. As you move the cursor, different boxes will highlight corresponding to the element that is accessible.

Figure 15.5:
Step 1: AND2 gate

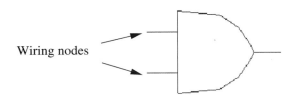

Wiring nodes

- Click and hold the left mouse button; then drag the cursor to the work area while continuing to depress the button. As soon as the cursor enters the work area, an outline of the component will appear.
- When the component is in the desired position, release the mouse button. This places it into the schematic.

A logic diagram is created by placing each component into the work area, then wiring them together using the Wire tool. This is activated by pushing the wiring button. Once you are in wiring mode, point to the first node you want to connect and hold the button down. Then drag the cursor to the other node; releasing the button will create an electrical connection between the two nodes.

Let us illustrate some fine points by means of an example. Suppose that we want to construct the logic diagram for the AND-OR (AO) function:

$$f = A \cdot B + C \tag{15.1}$$

Step 1: Drag an AND2 gate to the work area. The result is shown in Figure 15.5. The wiring nodes are defined at the ends of the wires, as indicated.

Step 2: Drag an OR2 gate to the work area. Place it to the right of the AND2 gate to allow wiring between the two. Figure 15.6 shows the schematic at this point.

Figure 15.6:
Step 2: Adding an
OR2 gate

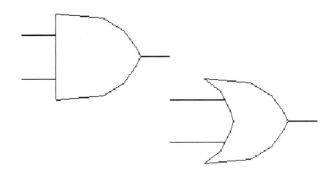

Figure 15.7:
Step 3: Wiring the gates

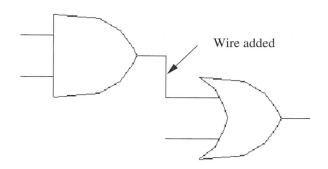

Wire added

Step 3:   Use the wiring tool to connect the output of the AND2 with an OR2 input. Activate the tool, click and hold on one node, drag the cursor to the other node, and then release the mouse button. The results of this operation are shown in Figure 15.7.

Step 4:   Although the schematic is completed, we need to add inputs and an output indicator to test it. We will use the push-button switches for entering values, and an LED at the output. Select a push-button switch, and drag it to the top input of the AND2 gate; align the left side of the box to the gate input and release the button. If you were close to the node, it will connect automatically to the gate. If not, you can wire them together. Do the same for the other inputs, and add an LED at the output. The final circuit is shown in Figure 15.8. The inputs and outputs have pre-assigned names (in1, in2, in3, and out1); these can be changed if desired, but we will leave them as is for now.

It is worthwhile to take the time to construct the schematic as detailed in the step-by-step sequence. You will find that the procedure is straightforward, even easy! The drawing skill will be useful when we progress to more complex functions.

Figure 15.8:
Step 4: Adding input and output components

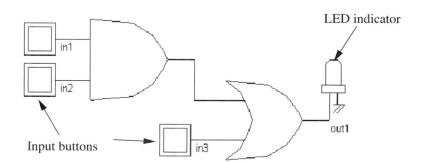

LED indicator

Input buttons

out1

● ● ● ● ● ● ● ● ● ● ● ● ● ⋯⋯

15.4 Simulating a Logic Design

Simulation is the best means for design verification. It is inexpensive (compared to fabricating a silicon chip!), and the results can be studied and analyzed for any problems. At the schematic level, Dsch allows you to play with the inputs and view the outputs manually. The results also can be viewed in timing charts.

Once a logic schematic is finished, the design is simulated by using the Simulate button shown here. Alternately, you may use the Menu command sequence:

Simulate ⇒

    Start Simulation

illustrated in Figure 15.9(a), or the keyboard combination **[Ctrl][S]**. This activates the circuit and displays the Simulation Control window [see Figure 15.9(b)] in the lower-left section of the Work area. When the simulation mode is activated, you can point and click on the input buttons. When a button is transparent, it is in a logic 0 state. Clicking on it will change it to red, which indicates a logic 1 state. As you click different buttons, the output LED indicator will also change to logic 0 (off) and logic 1 (on) states. If the **Show wire state** box is checked (as in the screen dump example), the internal wires will also change. In the color mode, a 0 state is blue while a red line indicates a logic 1. Once you have tested the desired combinations, you can exit the simulation mode by either closing the window or using the Menu command **Stop Simulation**.

Since the simulation described was performed in real time, you may not remember every input combination. A nice feature of Dsch is that your test session is recorded, and the results can be viewed using the Menu commands:

View ⇒

    Timing Diagrams

(a) Simulate menu   (b) Simulation Control window

Figure 15.9: Simulate menu and Simulation Control window

or by using the **View Timing** button shown here. It is worthwhile to take a moment and study the other commands in the View command window.

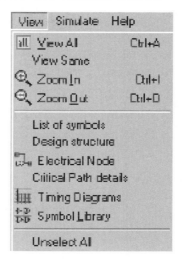

The timing diagrams are graphical records of the circuit stimulus and response that was recorded during the simulation session. If you rerun the simulation, the output will change. An example of a set of timing diagrams is shown in Figure 15.10 for the AND–OR circuit designed in the AO circuit example. The inputs (in1, in2, in3) are at 0 levels until the buttons are pushed. When this occurs, they switch to logic 1s. As each input changes, the output (out1) is updated to give the result. **Cursor** control helps you to select specific times and/or input combinations. The duration of the plots varies with the simulation session.

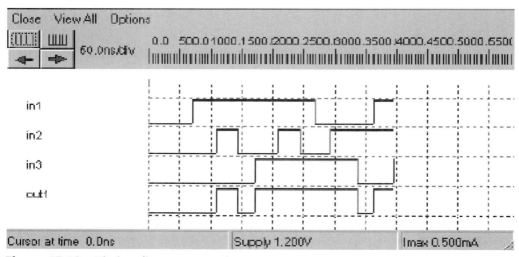

Figure 15.10: Timing diagram example

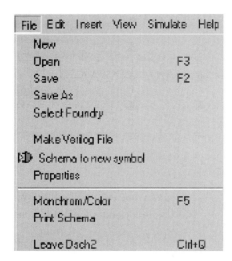

15.5 Creating a Macro Symbol

Hierarchical design relies on creating complex cells from simpler ones. At the system design level, it is convenient to take a network and define it as a macro for use in more complex designs. An example might be an adder, which is made up of primitive logic gates, but is thought of as a single unit.

Dsch allows you to create a macro symbol for any logic circuit and then embed the symbol into another design. Once a drawing has been completed, the symbol creation process is activated by the **Menu** command sequence:

File ⇒
    Schema to New Symbol

or by pushing the button shown here. The **File** window is reproduced in Figure 15.11. Note that several other commands are also available under the File heading. This operation opens a window that allows you to create a symbol for the logic network that is on the screen.

We will use the AO network from Figure 15.8 for our discussion. The **Schema to Symbol** dialog window for this circuit is shown in Figure 15.12. This automatically detects the input and output lines and uses them as ports. A listing of these is provided on the left side of the window. A **Symbol** preview is shown on the right side of the screen. You can increase or decrease the width of the symbol using the button controls. The **Symbol Properties** area allows you to define the Name and Title of the element. When the window is first opened, the default entries will be taken from the file name. These can be

Figure 15.12:
The Schema to Symbol dialog window

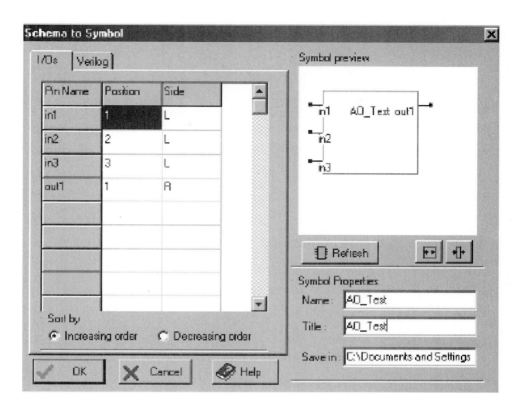

changed as desired; pushing the **Refresh** button will modify the symbol to the new name. Once you are satisfied with the overall specifications, it may be saved by pushing the OK button. We have named the symbol *AO_Test*.

Creating a macro symbol for the circuit allows you to insert it as needed. In other words, the symbol is a library cell entry that can be **instanced** (copied) as needed. To use this feature, use the Menu command:

Insert ⇒

and select the cell. An example is provided in Figure 15.13. The cell that was created for the AO circuit is the last entry in the window; the specific path listed will be different, depending upon the location of your data. You can also select other symbols as desired.

When a symbol is selected, an outline appears; the placement is controlled by the cursor, with the final location being the point where you click the mouse button. In the example portrayed in Figure 15.14, the symbol has been placed to the right of the original circuit. The symbol has the same logic characteristics as the original network but is in a more compact form. It may be used as a primitive element to create large logic units.

Figure 15.13:
Insert window

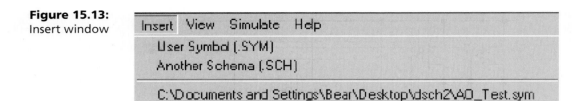

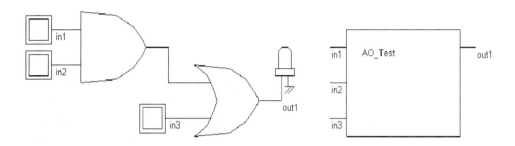

Figure 15.14: Insert symbol example

Another example is shown in the circuit of Figure 15.15. The AO symbol has been instanced into the design with a negative-edge triggered DFF. A clock is used to control the latching, and input buttons have been added to the AO cell. The outputs are monitored with LEDs. Running a simulation on this verifies that the *AO_Test* symbol has the same logic output as the primitive circuit.

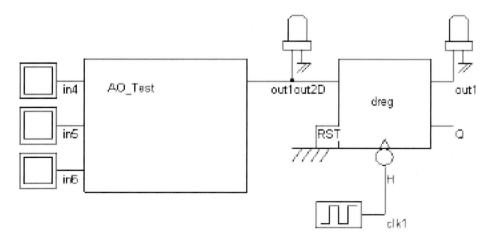

Figure 15.15: Logic circuit with instanced *AO_Test* symbol

Although we prefer to do system design from the top–down, the bottom–up approach illustrated here is used to create high-level complex logic functions. This is one of the keys to using hierarchical design. Even the most complex microprocessor is made up of primitive AND, OR, and NOT functions, but designing a 64-bit pipelined system would be difficult, if not impossible, if we were restricted to primitive logic gates.

● ● ● ● ● ● ● ● ● ● ● ● ● ● ● ● ●

15.6   Creating a Verilog® Listing

Hardware description languages (**HDLs**) allow you to describe a digital system using a text file. Gates and wiring are specified using a well-defined syntax and "dictionary-like" translation in a manner that one can construct the logic diagram entirely from the file. HDLs are used to test and verify a design before it is implemented in hardware. They contain specific information on delays and other hardware-specific parameters that allow detailed validation of the design in a particular process. In VLSI, most of the testing and verification is performed using an HDL.

Verilog HDL commonly is used in designing digital CMOS integrated circuits. It has several features that make it attractive to the chip designer including: FET models and the ability to test CMOS logic. Verilog is a free-flowing language, and its rules of construction are easy to learn. Many chip designers find that Verilog is written in a manner that is very close to the way they think. Its rules of construction are not as rigid as **VHDL**, which is the other dominant HDL used in digital system and chip design.

Although we will examine some aspects of a Verilog network description, we will not present the details of Verilog coding here The interested reader is referred to one of the texts listed at the end of this chapter. Our emphasis will be on using the generated Verilog listing as a means of transferring the system design information to different levels.

The Dsch program can take a logic circuit and translate it into a Verilog description that uses a slightly simplified syntax. The core of the Verilog file is equivalent to a netlist that describes the components and how they are connected. This feature is activated using the Menu command sequence:

File ⇒
    Make Verilog File

as seen in the listing of Figure 15.11. When applied to the AO circuit of Figure 15.8, this action opens up the window displayed in Figure 15.16. The Verilog listing is shown on the left side. Information about the listing, including the number of lines, symbols (components), and nodes is shown on the right side of the screen.

The Verilog description is saved as a text file in the Dsch folder. It is reproduced here so that we can investigate some of the important properties. In particular, it is important to point out some differences between the simplified coding produced by Dsch and the standard Verilog syntax. The Dsch Verilog file is written specifically to interface with Microwind so that the differences do not affect the design flow.

Figure 15.16:
Verilog file for
AO circuit

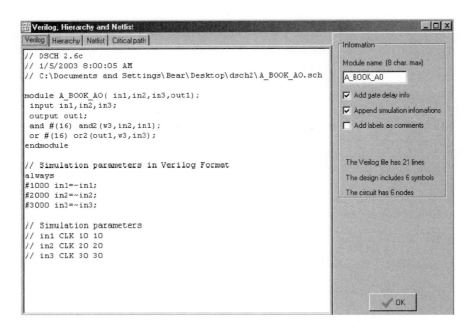

In the listing below, a double front slash "//", indicates a comment line.

// DSCH 2.6c
// 1/5/2003 11:00:08 AM
// C:\Documents and Settings\Bear\Desktop\dsch2\A_BOOK_AO.sch

module A_BOOK_AO(in1,in2,in3,out1);
    input in1,in2,in3;
    output out1;
    and #(16) and2(w3,in2,in1);
    or #(16) or2(out1,w3,in3);
endmodule

// Simulation parameters in Verilog Format
always
#1000 in1=~in1;
#2000 in2=~in2;
#3000 in3=~in3;

// Simulation parameters
// in1 CLK 10 10
// in2 CLK 20 20
// in3 CLK 30 30

The first three lines are just informational comment lines. The main portion of the Verilog description is in **module** definition. This has the form:

module A_BOOK_AO(in1,in2,in3,out1);

…

endmodule

where the listing is inserted between the lines. A module may be thought of as a cell with inputs, outputs, and controls. The first line is a declaration statement. In this case, it defines a module named *A_BOOK_AO* with input and output identifiers in parentheses: (in1,in2,in3,out1). The generated file lists the inputs first and the output last; this is opposite to standard Verilog syntax. The next lines declare which of the identifiers are inputs (in1,in2,in3) and outputs (out1). Verilog uses semicolons as delimiters on most statements.

The component list follows. The listing

and #(16) and2(w3,in2,in1);
or #(16) or2(out1,w3,in3);

uses the syntax

Primitive #(time_delay) Gate_Name(output, input1, input2,...);

where

- Primitive identifies the function (and, or)
- time_delay is the delay in relative (normalized) units
- Gate_Name is optional, but used to keep track of components

We therefore interpret the code as describing two gates, both of which have a delay of 16 time units. An important comment is that the delay Dsch uses, depends upon the technology available through the **Select Foundry** command. The two gates have names of and2 and or2, and are implied to be 2-input since there are only two input variables listed in each. Note that a new identifier, w3, appears in the gate listing. This is called a **wire** in Verilog and represents an identifier for a line that is used as an internal interconnect. In this example, w3 is taken from the output of the and2 gate and is used as an input to the or2 gate. In standard Verilog syntax, wires are declared using a statement such as

wire w3;

However, Microwind automatically interprets wires when it reads the Verilog file, so that the declaration is not used here. The circuit can be reconstructed from the module listing by:

1. Drawing the gates with inputs and output labelled

2. Connecting nodes with the same name

3. Wiring the input and output port identifiers to the internal wiring

That completes the description of the module, which is the important information on the elements and how they are connected.

The remaining lines in the Verilog listing are for inputs. The statement

#1000 in1=~in1;

takes the input in1 and changes it to ~in1 (~ is the NOT operation) every 1000 time units. In simple terms, this is a clock with a period of 2000 units. Stimuli are not part of the module listing and are used for simulation purposes.

The important point of this section is that the schematic can be translated into an HDL code for use in other parts of the design process.

15.6.1   Compiling the Verilog Code

To complete the discussion, we will now minimize Dsch and launch the Microwind program. This will allow us to take the design down to the physical level using the CAD tools.

The Verilog listing generated by Dsch can be compiled into a CMOS circuit using Microwind. Executing the **Menu** command sequence:

Compile ⇒
   Compile Verilog File

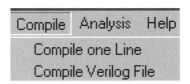

opens a window listing the file. The Verilog file will be in the Dsch folder (unless you changed the path). Opening the folder will list all Verilog files. After finding the file of interest and clicking on it, open the dialog window in Figure 15.17, where we have selected the AO listing discussed above. The upper window shows the Verilog description, while the right side of the screen provides some options.

Microwind will translate this into a CMOS circuit by hitting the **Compile** button on the lower-left side of the screen. A log is created in the lower (originally empty) screen to document the process. The log for this example has been reproduced in Figure 15.18 to show some of the detail. Closing this window and returning to the editing screen displays

Figure 15.17:
Microwind Compile
Verilog File window

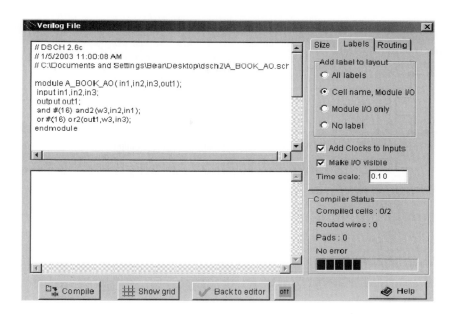

the layout for the CMOS circuit in Figure 15.19. The power of the procedure immediately becomes obvious. The CAD tools allow one to take a design from logic schematic directly to a physical CMOS circuit with a straightforward procedure! This allows us to implement very complex designs without having to go through manual layout. The main drawback of the process is that the CMOS circuit will probably not be the smallest possible, nor will it be optimized for speed. That usually requires that we refine the layout or even do it over from scratch.

Figure 15.18:
Compiler log in
Microwind

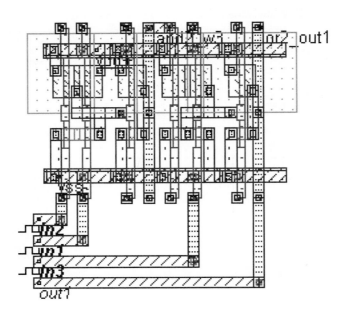

● ● ● ● ● ● ● ● ● ● ● ● ● ●

15.7 The Dsch–Microwind Design Flow

The example in this chapter provides a step-by-step approach to high-level digital design. This is summarized in Figure 15.20. Logic design and validation is performed in Dsch. When that is completed, a Verilog file is generated and sent to Microwind where it can be compiled into a CMOS layout. This allows the designer to test the circuit operation using SPICE simulations.

● ● ● ● ● ● ● ● ● ● ● ● ● ●

15.8 Using a Design Toolset

Dsch and Microwind are very powerful programs. However, in some cases, the fully automated design flow may not be able to produce a finished CMOS circuit. Sometimes, it is a limitation of the program: other times, the problem arises from the user. Regardless, it is always important to understand the limitations of any CAD tool that you use and how to avoid, work around, or overcome them.

Figure 15.20:
Dsch–Microwind
design flow

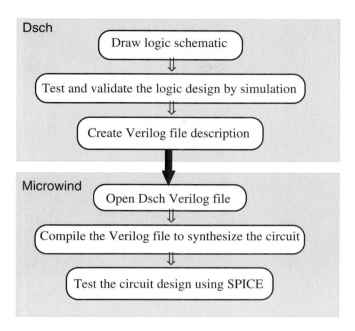

In this example, we will use a generic 3-input complex gate from the Dsch Symbol library to define modules for a full-adder network. The complex gate is identified by the button shown here. Note that there is also a 5-input gate available. Dragging this into the work area produces the symbol shown in Figure 15.21. The default logic expression is

$$s \; = \; a|b|c \qquad\qquad (15.2)$$

which is the OR3. However, the logic can be changed by the user. Double-clicking on the symbol opens the dialog window in Figure 15.22. We have changed the expression to read as

Figure 15.21:
Initial appearance
of the complex
logic gate symbol

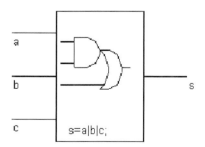

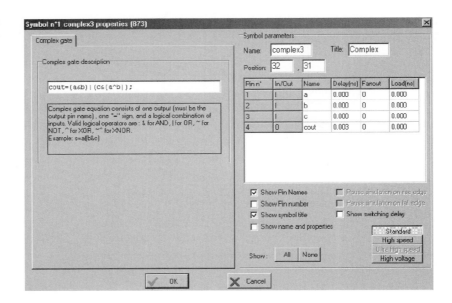

$$cout = (a \ \& \ b) \ | \ (c \ \& \ (a \wedge b))$$

which is equivalent to the more familiar expression:

$$c_{\text{out}} = a \cdot b + c \cdot (a \oplus b) \tag{15.3}$$

for the carry-out bit of a full adder. To make the ports agree, we also change the name from "s" to "cout" on the right-side port listing. Clicking OK returns you to the work screen, and the symbol will have the new expression.

Next, we instance another 3-input complex logic gate and redefine it to have an output of

$$\text{sum} = a \wedge b \wedge c$$

or

$$\text{sum} = a \oplus b \oplus c \tag{15.4}$$

for the sum bit. Arranging the gates and adding input buttons and output LEDs gives the circuit shown in Figure 15.23. Running a simulation on this circuit verifies that it is a valid full-adder network, so that the design flow is valid up to this point.

To translate this into a CMOS circuit, we create and save a Verilog file, then open it in Microwind. However, the Compile function experiences an error and is not able to produce the circuit.[1] To find the problem, let us examine the Verilog listing created by Dsch:

---

1. This problem occurred in the Microwind2 version available when this book was written.

Figure 15.23:
Full adder using
configurable
3-input logic gates

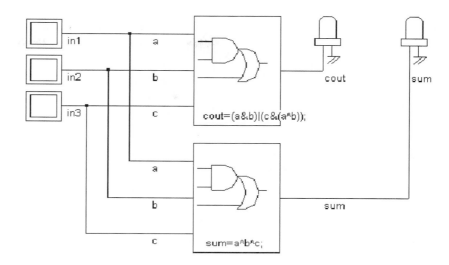

```
module A_BOOK_ComplexFA(in1,in2,in3,sum,cout);
    input in1,in2,in3;
    output sum,cout;
    assign cout=(in1&in2)|(in3&(in1^in2));
    assign sum=in1^in2^in3;
endmodule
```

This uses the Verilog statement **assign** which is classified as a high-level abstract command. It commonly is used to describe systems at the **register–transfer level (RTL)** and generally does not provide much information as to the interior of the cell itself.

The error listing specified that it has a problem with Line 5 and the assign statement. Yet, it was able to compile the circuit for the carry-out bit, cout, so the problem must be in the statement:

assign sum=in1^in2^in3;

Although this is a valid Verilog statement, and the Dsch simulations worked fine, the Microwind Compile operation tends to work best when parentheses are used to pair variables in expressions with three or more inputs. If this is changed to

assign sum=(in1^in2)^in3;

by adding parenthesis to the Verilog file, Microwind is able to compile the function and produce the layout in Figure 15.24. Alternately, if you change the Dsch block definition to
sum = (a^b)^c
then everything works fine.

Figure 15.24:
Full-adder circuit compiled using complex gates

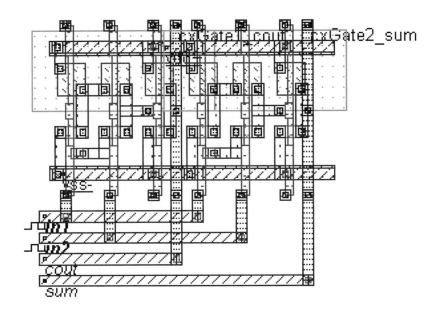

Although you may not experience this problem with the current software versions, it demonstrates an important point. Every CAD tool exhibits odd behavior at one time or another. A good designer *never* has a 100% confidence level in a computer result. If a tool does not work properly, then it could be something as simple as a syntax problem or indications of a major problem in the code. The best way to learn a program is to use it and find out any quirks as soon as possible.

● ● ● ● ● ● ● ● ● ● ● ● · · · ·

15.9 MOSFETs in Dsch

CMOS circuits can be built and simulated in Dsch using the nFET and pFET buttons in the **Symbol** library. They can be used to create a transistor-only network or can be merged with gate-level logic symbols.

A simple NOT gate is shown in Figure 15.25. This was constructed by dragging nFET and pFET symbols to the work area and then adding the power supply and ground connections. The input button and output LED were included to monitor the simulation. The drawing shows transistor aspect ratios of $(W/L)_n = (1.0 \ \mu m/0.12 \ \mu m)$ and $(W/L)_p = (2.0 \ \mu m/0.12 \ \mu m)$; the channel length is in the default 0.12 μm technology. Dsch allows the user to specify different aspect ratios as needed. Double-clicking on a FET symbol opens

Figure 15.25:
Inverter circuit
in Dsch

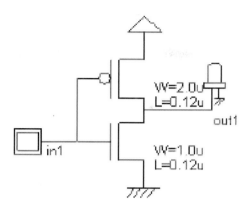

the dialog window shown in Figure 15.26. Device sizes can be entered on the left side of the screen. The data on the right side provides control over parameters, such as the name and what is to be displayed with the device.

The circuit may be tested using the **Simulation** command. When FETs are included in the circuit, Dsch changes them to switches to clearly illustrate their behavior with different inputs. Many people will find this an outstanding feature for learning the operation of CMOS networks. Figure 15.27 provides screen dumps from the simulation of the inverter. In Figure 15.27(a), the input is a logic 0. This closes the pMOS switch and opens the nMOS switch, producing a logic 1 (lit LED) at the output. Figure 15.27(b) has a

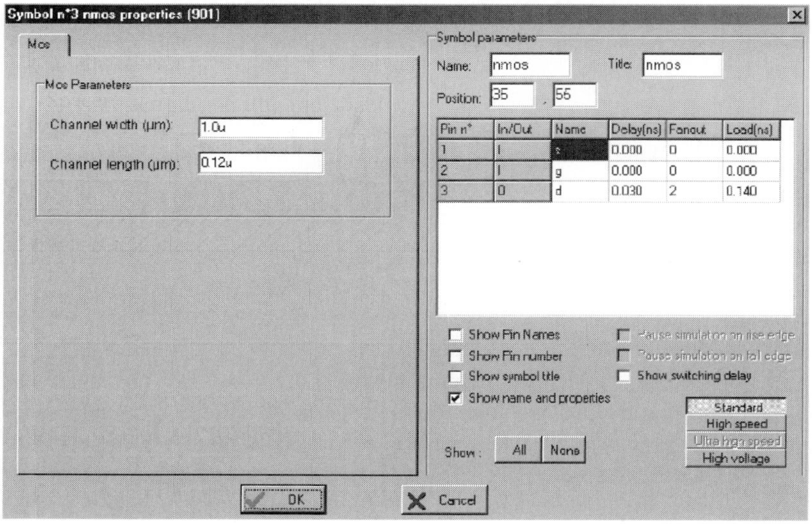

Figure 15.26:
FET dialog window

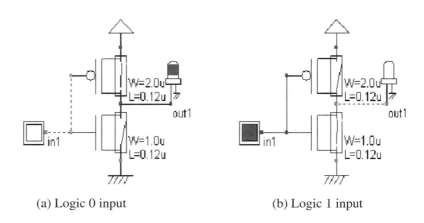

(a) Logic 0 input   (b) Logic 1 input

logic 1 input, which reverses the switching states. The data can be viewed as waveforms, as with the other Dsch networks.

Verilog provides nMOS and pMOS devices as primitive components. These can be seen by creating the Verilog file for this circuit. This results in the module listing below.

```
module NOT_CIR(in1,out1);
    input in1;
    output out1;
    nmos #(17) nmos(out1,vss,in1); // 1.0u 0.12u
    pmos #(17) pmos(out1,vdd,in1); // 2.0u 0.12u
endmodule
```

The syntax is straightforward. The keywords, nmos and pmos, specify the component. The pound sign (#) is used to specify delays, and the names, nmos and pmos, preceding the input/output list are arbitrary. The port list places the gate terminal as the last entry. The device sizes are specified as comments, as these are not part of a standard Verilog description.

The CMOS circuit can be generated by launching Microwind and using the **Compile Verilog** directive. A unique feature of this process is that the FET sizing information is used to produce the layout shown in Figure 15.28. The locations of the transistors have been added to the screen dump. The compiled circuit creates the $W_n = 1.0$ μm nFET as a single device but uses two pFETs in parallel to achieve $W_p = 2.0$ μm. This is not an efficient layout, but it is a valid one. Automated chip design flows often exhibit a trade-off between quick turnaround and efficient solutions.

Figure 15.28:
Compiled inverter
circuit

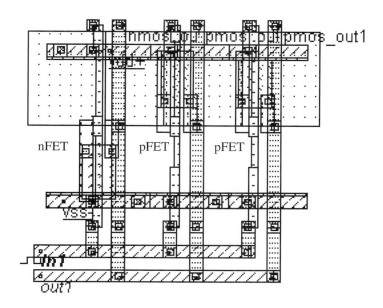

An a second example, let us create a NOR2 gate. The schematic is shown in Figure 15.29. The Verilog module is

```
module NOR2_Cir(in1,in2,out1);
    input in1,in2;
    output out1;
    nmos #(24) nmos(out1,vss,in2); // 1.0u 0.12u
    nmos #(24) nmos(out1,vss,in1); // 1.0u 0.12u
    pmos #(24) pmos(out1,w4,in1); // 2.0u 0.12u
    pmos #(10) pmos(w4,vdd,in2); // 2.0u 0.12u
endmodule
```

where it is seen that the nMOS transistors are in parallel (same node listing), while the pFETs are in series using the wire, w4. Compiling the Verilog file using Microwind yields the CMOS layout in Figure 15.30. This solution uses two nFETs and two pairs of pFETs.

Although Verilog nMOS and pMOS primitives represent the basic switching characteristics of n- and p-channel MOSFETs, it should be mentioned that they do not behave the same. One important difference is that the Verilog switches are uni-directional, whereas real transistors can conduct current in either direction. If you use nMOS and pMOS devices in Dsch, you should always simulate the circuit to ensure that it acts as desired. If

Figure 15.29:
NOR2 circuit

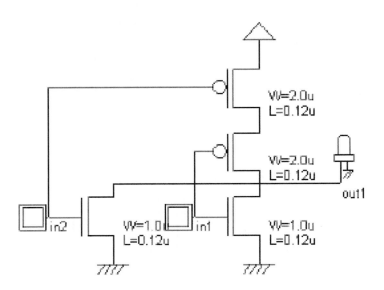

a switch is not transmitting, flip it to reverse the input and output (source and drain) terminals and try again.

The two orientations are shown in Figure 15.31. Proper operation is observed for the circuit in Figure 15.31(a). However, if the transistor is flipped, as in Figure 15.31(b), it will not pass the input. The Verilog listing for the two configurations is

Figure 15.30:
Compiled NOR2
circuit layout

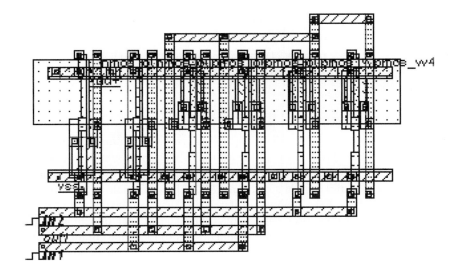

Figure 15.31:
Pass transistor
orientations for
Verilog devices

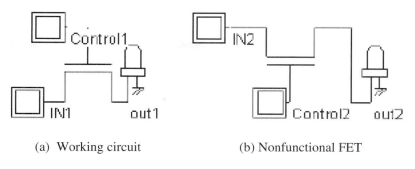

(a) Working circuit   (b) Nonfunctional FET

module PassFET(IN1,Control2,IN2,Control1,out1,out2);

    input IN1,Control2,IN2,Control1;

    output out1,out2;

    nmos #(10) nmos(out1,IN1,w2); // 1.0u 0.12u

    nmos #(10) nmos(IN2,out2,Control2); // 1.0u 0.12u

endmodule

If this is compiled by Microwind, it produces the layout shown in Figure 15.32. At this point, the FETs are physical devices, not Verilog constructs. Applying sources and simulating shows that the transistors indeed can pass signals in both directions!

Figure 15.32:
Compiled pass
transistor layout

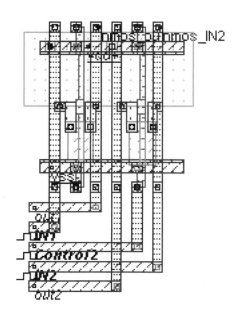

For simple circuits, Dsch may be used as a schematic capture tool. However, it is important to remember that the Microwind compiled circuit will not be as compact as one drawn manually. In addition, the node capacitances will be larger due to the additional interconnect lines, so that the circuit will also exhibit slower responses. That being said, the automatic compilation feature using the Verilog link between the two programs is a very powerful feature that permits a rapid translation from logic schematic to a CMOS circuit layout.

● ● ● ● ● ● ● ● ● ● ● ● ● ● ●

15.10 References

[15.1] Ciletti, M. D., *Advanced Digital Design with the Verilog HDL*. Upper Saddle River, NJ: Prentice Hall, 2003.

[15.2] Smith, M. J. S., *Application-Specific Integrated Circuits*. Reading, MA: Addison-Wesley, 1997.

[15.3] Uyemura, J. P., *Introduction to VLSI Circuits and Systems*. New York: John Wiley & Sons, 2002.

● ● ● ● ● ● ● ● ● ● ● ● ● ● ●

15.11 Exercises

15.1 Use Dsch to draw the logic diagram for the function:

$$F = a \cdot (b + c)$$

Simulate the circuit and check the logic. Then use Microwind to create the CMOS circuit.

15.2 Design the logic circuit for the function:

$$f = \overline{a \cdot b \cdot c + d \cdot e}$$

and then use Microwind to compile the CMOS circuit.

15.3 Design the logic circuit for the function

$$g = \overline{a \cdot b + a \cdot (b + c)}$$

Compile this using Microwind. Is this the best implementation?

Digital System Design 2
Design Flow Examples

In the last chapter, we studied the automated design flow provided by the Dsch and Microwind programs. This chapter provides some examples of system development using this approach. It takes the discussion from basic circuits to more complex networks as a means to illustrate the concepts of repetition and regularity in layout.

16.1 A 4-bit Binary Adder

Binary adders are often used as test circuits, because they provide moderate complexity in addition to being important modules in digital systems. We have already studied the full adder, but this time we will invoke the power of the automated process to see how it can be used to build a 4-bit adder unit.

16.1.1 The 1-bit FA Cell

First we define the FA module, as shown in Figure 16.1. This calculates the binary sum, s_n of the input bits: a_n, b_n, and the carry-in bit c_n, from

$$s = a_n \oplus b_n \oplus c_n \tag{16.1}$$

The carry-out bit is given by

$$c_{n+1} = a_n \cdot b_n + c_n \cdot (a_n \oplus b_n) \tag{16.2}$$

Figure 16.1:
Full-adder module

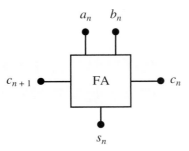

These equations are used to draw the gate-level Dsch circuit shown in Figure 16.2. Note that a direct translation of the expressions has been used. A simulation yielded the timing diagrams in Figure 16.3, which verifies the operation.

The next step is to create a symbol for the circuit that can be instanced into other designs. Using the **Menu** command sequence:

File ⇒

Schema to New Symbol

Figure 16.2:
Gate-level schematic for the full-adder module

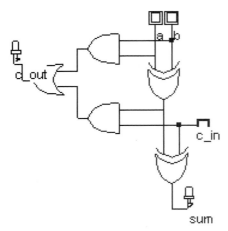

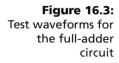

Figure 16.3:
Test waveforms for
the full-adder
circuit

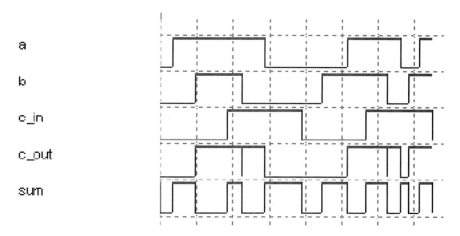

opens the symbol definition window. The default design has inputs on the left and outputs on the right. For this example, the symbol will be modified to that shown in Figure 16.4, as this gives easy wiring for a 3-bit circuit. Ports can be moved by specifying the side:

- L = left
- R = right
- T = top
- B = bottom

Figure 16.4:
Symbol definition
window

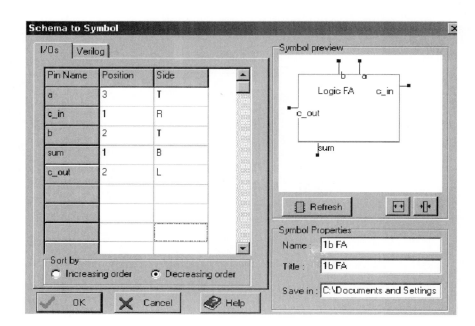

This gives the convenient structure displayed on the right side. We have named the symbol "1b FA". Pushing the **Refresh** button updates the symbol to display the new orientation. When all modifications are finished, the **OK** button saves the symbol in the Dsch folder.

The next step in the process is to create the Verilog description. This produces the module listing:

```
module Logic FA(a,c_in,b,sum,c_out);
    input a,c_in,b;
    output sum,c_out;
    and #(16) and2(w3,w1,c_in);
    xor #(23) xor2(w1,b,a);
    xor #(16) xor2(sum,c_in,w1);
    and #(16) and2(w7,b,a);
    or #(16) or2(c_out,w3,w7);
endmodule
```

which is a gate-by-gate description of the circuit.

To obtain a CMOS layout, we launch Microwind and execute the **Compile Verilog** command and specify the file. The generated layout is shown in Figure 16.5. An option on the **Compile Verilog** dialog window is **Add Vertical Bus** wiring. If this is selected, the compiled circuit has vertical data lines, as shown in Figure 16.6. This may be useful, depending upon the desired shape of the final 4-bit circuit.

Figure 16.5:
Compiled 1-bit full-adder CMOS circuit

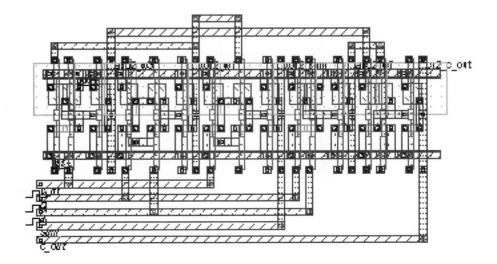

Figure 16.6:
1-bit full adder with
vertical I/O lines

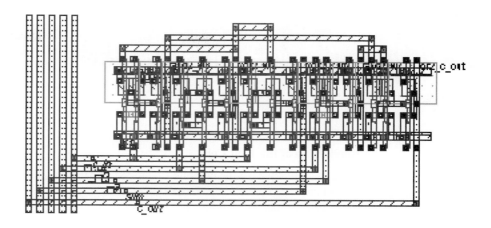

16.1.2   Building the 4-bit Adder

Now that we have a 1-bit FA cell, it can be used to build a 4-bit adder. The simplest approach is to define two 4-bit words:

$$a = a_3 a_2 a_1 a_0$$
$$b = b_3 b_2 b_1 b_0$$

(16.3)

and construct the 4-bit **ripple carry adder** shown in Figure 16.7. This uses the carry bit, c_n, as an input to the $(n+1)$ cell so that the carry bits "ripple" from bit 0 towards bit 3. This can be implemented in two ways.

Figure 16.7:
4-bit ripple
carry adder

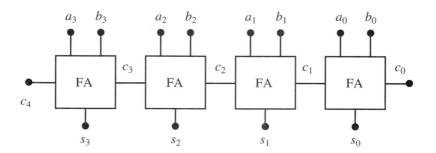

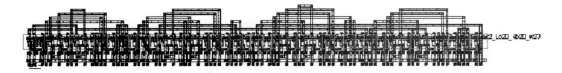

Figure 16.8: 4-bit adder from replication of a 1-bit cell

Replication of the 1-bit Cell

Since we already have the complied 1-bit cell, it can be replicated to create a 4-bit adder. The Microwind command is **Duplicate** X,Y. This result is shown in Figure 16.8; you easily can see that the primitive has been copied to produce the 4-bit adder in a row geometry, with X = 4 and Y = 1. An alternate orientation is shown in Figure 16.9. This circuit is the same 4-bit adder, but is created using the command X = 1 and Y = 4, so that the result is a column. Note that this provides wiring I/O lines to be run from the sides, and is automatically structured for an 8-bit segment by duplicating the column and flipping it horizonnatally. Another possibility is building a row of cells with the primitive cells accessed by vertical wiring. This is shown in Figure 16.10 and may be useful in situations where the overall wiring is along vertical channels instead of the horizontal channels used in the

Figure 16.9:
4-bit adder with
column structure

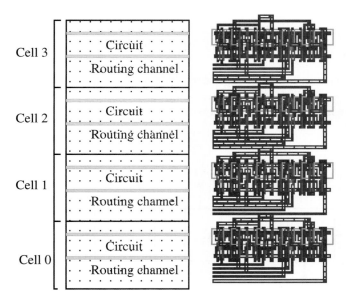

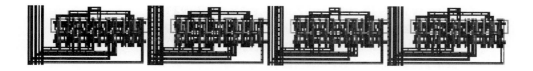

Figure 16.10: 4-bit adder with vertical wiring paths

other examples. Since the vertical wiring is at a higher level metal than the horizontal input and output lines, the general interconnect rules are satisfied.

Compilation of a 4-bit Logical Adder

The alternative to replicating a 1-bit cell is to design and compile the 4-bit adder as a single symbol. Using the **Insert** command, the 1-bit FA cell may be instanced to create the Dsch circuit shown in Figure 16.11. This has been wired according to the original specification in Figure 16.7, so that there is a direct one-to-one correlation. Hex keyboards have been added to verify the design. Once the design is validated, we will create the more general network shown in Figure 16.12; this is not necessary, but makes the compilation a little easier. We will first create the symbol using the dialog window in Figure 16.13. The symbol itself becomes a single entity, and only the overall macro characteristics are important. The Verilog file can be generated in the form

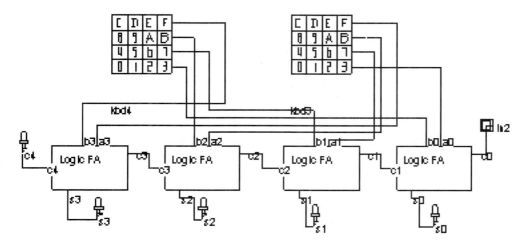

Figure 16.11: 4-bit adder circuit using 1-bit FA cells

```
module 4bAdder( );
    wire w14,w15,w16,w17,w18,w19,w20,w21;
    wire w22,w23,w24,w25,w26,w27,w28,w29;
    and #(15) and2_Lo1_4b1(w16,w14,w15);
    xor #(27) xor2_Lo2_4b2(w14,w1,w2);
    xor #(7) xor2_Lo3_4b3(w13,w15,w14);
    and #(15) and2_Lo4_4b4(w17,w1,w2);
    or #(25) or2_Lo5_4b5(w18,w16,w17);
    and #(15) and2_Lo6_4b6(w20,w19,w18);
    xor #(27) xor2_Lo7_4b7(w19,w6,w5);
    xor #(7) xor2_Lo8_4b8(w9,w18,w19);
    and #(15) and2_Lo9_4b9(w21,w6,w5);
    or #(7) or2_Lo10_4b10(w10,w20,w21);
    and #(15) and2_Lo11_4b11(w24,w22,w23);
    xor #(27) xor2_Lo12_4b12(w22,w8,w7);
    xor #(7) xor2_Lo13_4b13(w12,w23,w22);
    and #(15) and2_Lo14_4b14(w25,w8,w7);
    or #(25) or2_Lo15_4b15(w15,w24,w25);
    and #(15) and2_Lo16_4b16(w28,w26,w27);
    xor #(27) xor2_Lo17_4b17(w26,w3,w4);
    xor #(7) xor2_Lo18_4b18(w11,w27,w26);
    and #(15) and2_Lo19_4b19(w29,w3,w4);
    or #(25) or2_Lo20_4b20(w23,w28,w29);
endmodule
```

This shows that the 1-bit adder circuit has been duplicated four times with the bit positions indicated by the change in variable labeling. To create the CMOS circuit, this file is opened by the **Compile Verilog** command in Microwind. Execution yields the circuit in Figure 16.14. The input wiring channel is seen clearly below the VSS line, illustrating

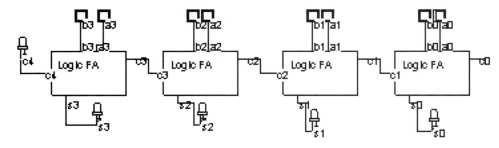

Figure 16.12: 4-bit ripple carry adder with standard I/O

Figure 16.13:
Symbol design for
the 4-bit adder

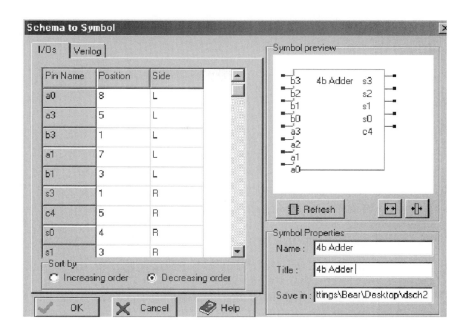

the difference between the circuit compiled as a unit versus that created by replication. Which one is better? It depends upon how the overall structure fits on the chip. Some geometrical shapes will be better than others.

16.2   Carry Look-Ahead Adder

The latency associated with the ripple carry adder can become a problem in a high-speed design. The carry look-ahead adder (CLA) overcomes this by rewriting the carry-out bit in the form:

$$c_{n+1} = g_n + p_n \cdot c_n \qquad (16.4)$$

Figure 16.14:
Fully compiled 4-bit
adder circuit

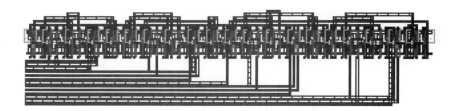

where

$$g_n = a_n \cdot b_n \tag{16.5}$$

is called the **generate** term, and

$$p_n = a_n \oplus b_n \tag{16.6}$$

is the **propagate** term. Note that the sum bit, s_n, is calculated from

$$s_n = p_n \oplus c_n \tag{16.7}$$

In general, the AND and XOR functions may be used to show that

$$g_n \cdot p_n = 0 \tag{16.8}$$

so that only the generate or propagate term is 1 for any input combination.

The structure of a CLA adder is shown in Figure 16.15. The inputs are used to find the generate and propagate terms, which in turn, are used to calculate the carry bits. Once the carry bits are computed, the sum is found using XOR gates. We will concentrate on designing the carry calculation logic and circuit block.

The CLA carry-bit equations are nested. This is seen by writing them out explicitly:

$$
\begin{aligned}
c_1 &= g_0 + p_0 \cdot c_0 \\
c_2 &= g_1 + p_1 \cdot c_1 \\
c_3 &= g_2 + p_2 \cdot c_2 \\
c_4 &= g_3 + p_3 \cdot c_3
\end{aligned}
\tag{16.9}
$$

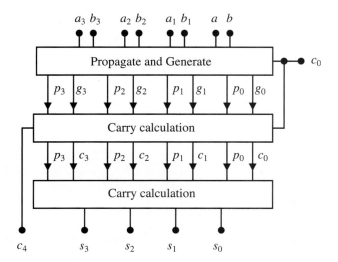

Figure 16.15: Architecture of a carry look-ahead adder

Figure 16.16:
CLA carry logic
circuit

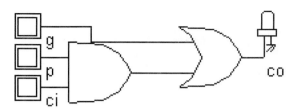

Since each expression has the same form and uses an input from the equation above it, we only have to design one circuit and then cascade copies. The basic unit, as constructed in Dsch, is displayed in Figure 16.16. This can be used to create the symbol in Figure 16.17. The inputs have been arranged so that they easily will interface in a cascade. A Verilog file can be generated and complied by Microwind resulting in the layout of Figure 16.18. This completes the design of the circuit for c_1.

The nested structure of the equation set, allows us to design the c_2 circuit using the c_1 files. The logic diagram is shown in Figure 16.19. This is the original c_1 schematic with an added CLA macro symbol; the symbol itself was created using the **copy** operation, and the labeling was modified in the Dialog window. This can be used to generate a Verilog description, which is then compiled by Microwind to produce the CMOS layout in Figure 16.20. A careful inspection of the signal routing below the VSS line shows that the inputs are in the same order as for the c_1 network, simplifying the overall design of the unit.

The c_3 and c_4 circuits are built using the same technique. Figure 16.21 shows the logic diagrams for the two carry bits. The nesting is obvious in the construction. The compiled CMOS layouts in Figure 16.22, maintain the same structuring and routing characteristics as the previous plots. The complete carry-calculation unit can be built by combining the four circuits into one and completing the interconnect wiring. This task (along with the testing and simulation of the circuits) is left as an exercise for the reader. Manual layout of CLA circuits can be tedious, but the automated design flow greatly simplifies the work. It is important to remember that the CMOS layout generally is not optimized in this

Figure 16.17:
CLA symbol created
in Dsch

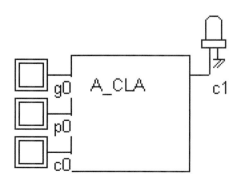

Figure 16.18:
CLA symbol created
in Dsch

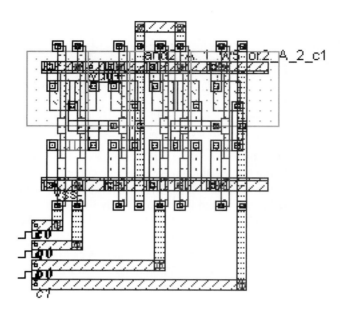

approach. However, the real test is whether the circuit performance is acceptable for the application at hand.

The objective of this example was to illustrate how the structure of logic equations can be used to simplify the design of the CMOS realization. Since it is possible to write a logic function in different forms using Boolean identities and DeMorgan operations, the

Figure 16.19:
Logic diagram for
the c_2 carry bit

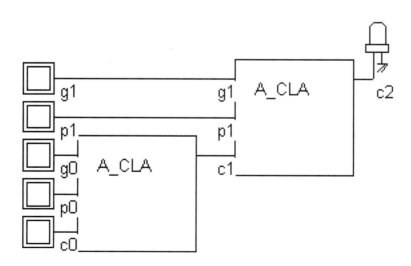

Figure 16.20:
CMOS circuit for
the c_2 bit

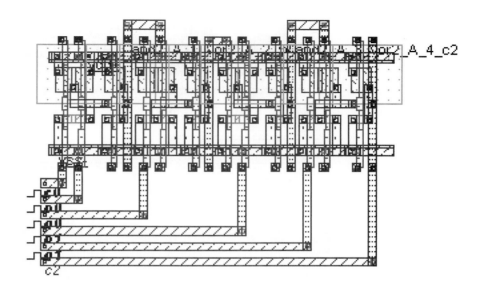

Figure 16.21:
Logic circuits for
the higher-order
carry bits

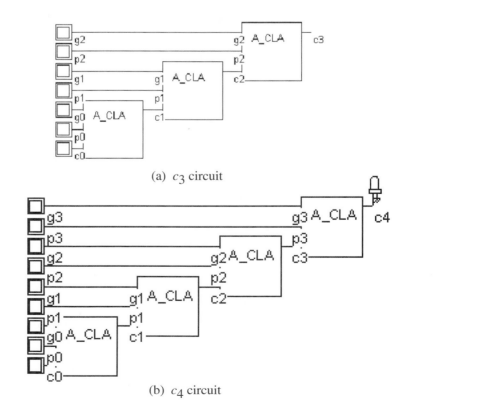

(a) c_3 circuit

(b) c_4 circuit

Figure 16.22:
Compiled CMOS
circuits for c_3 and c_4

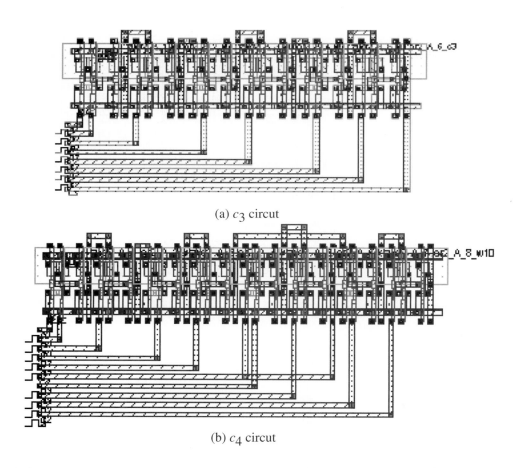

(a) c_3 circut

(b) c_4 circut

designer sometimes has the authority to restructure the circuits by starting at the logic level. The CLA design example also demonstrated the use of logic macros (through the definition of the CLA symbol) for simplifying the testing and simulation phase of VLSI design. General observations such as these, often help the designer in complex situations.

● ● ● ● ● ● ● ● ● ● ● ● ● ● ● ● ···

16.3 Pipeline Register

A register is a set of storage circuits that can hold an n-bit word. Pipeline registers are used to control the flow of data through each block of logic circuits.

Binary circuits deal with individual bits. To construct an n-bit register, we start with a 1-bit storage cell and replicate it to the desired width. Data flow through a pipeline is controlled with a clock signal, ϕ, so we will use edge-triggered DFFs. Figure 16.23 shows the development of a pipeline 8-bit register. Single-bit DFFs [Figure 16.23(a)] form the start-

Figure 16.23:
Development of an
8-bit pipeline
register

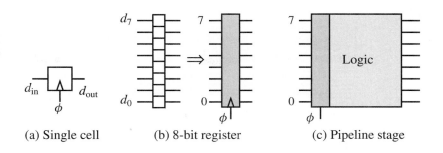

(a) Single cell　　　(b) 8-bit register　　　(c) Pipeline stage

ing point. Eight of these are paralleled to form the register, as in Figure 16.23(b). The locations are now numbered 0 through 7, and the simplified block symbol replaces the single-bit blocks. The register is then placed in front of a logic block to control the data flow into the circuits. With positive edge-triggered DFFs, data is admitted to the logic block on every rising clock signal.

Drawing the logic schematic in Dsch is simplified because a negative edge-triggered DFF is one of the primitive elements in the Symbol library. There are two obvious approaches to building an 8-bit register: create the CMOS circuit for a 1-bit DFF then replicate it, or design the 8-bit register in Dsch and compile the complete network. Since these are the obvious procedures, let's try a different approach. We will create a 4-bit register circuit and then replicate it to obtain the 8-bit design. This will illustrate the important features of the obvious routes while introducing some new ideas into the discussion.

To start, we will drag a single-bit DFF onto the work screen, add an input button and an LED monitor, and then replicate it to create the 4-bit circuit in Figure 16.24(a). We

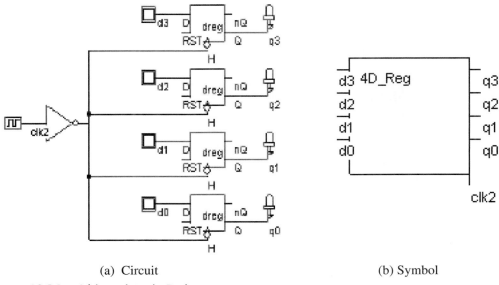

(a)　Circuit　　　　　　　　　　　　(b) Symbol

Figure 16.24:　4-bit register in Dsch

Figure 16.25:
CMOS circuit for
the 4-bit register

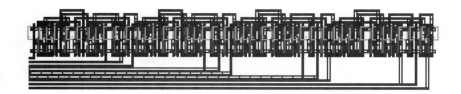

have added a clock and an inverter to obtain a circuit that triggers on positive edges. The inputs have been labelled d_3, d_2, d_1, and d_0: and the outputs are denoted as q_3, q_2, q_1, and q_0. The symbol in Figure 16.24(b) maintains the same notation.

The CMOS circuit is created with the standard procedure. First, the logic is translated into a Verilog listing in Dsch. This automatically saves the file to the Dsch folder, which can then be opened with the **Compile Verilog** command in Microwind. The resulting 4-bit CMOS layout is shown in Figure 16.25. A 1-bit section of the design is shown in Figure 16.26 for reference.

To create the 8-bit circuit, we will perform a **Copy** function on the 4-bit design. Following this with a **Flip Vertical** command and merging the clock lines together results in the final circuit shown in Figure 16.27. This approach allows us to do a vertical stacking, but it does require that the VDD and VSS lines be flipped, as in a Weinberger layout. Of course, many other structural variations are possible, even with the limitations imposed by the compiled circuit layouts.

This example illustrates some of the interaction between the structure of basic cells and how they affect the final layout in the automated flow. In most cases, the chip designer must satisfy a myriad of constraints and specifications. Some of these will revolve around layout and placement problems, so that it is advantageous to expand one's horizons by thinking about different techniques.

Figure 16.26:
1-bit section of the
D-register design

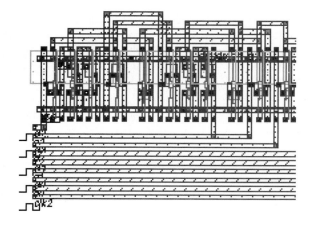

Figure 16.27:
Final layout of the
8-bit register using
a flipped cell

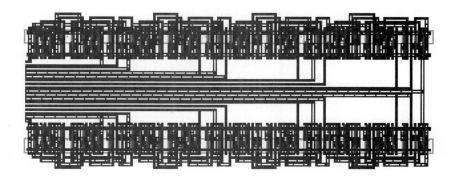

16.4   Divide-by-*N* Circuit

A divide-by-*N* circuit takes a clock with frequency, *f*, and period, *T*, and produces an output that has a frequency of *f*/*N* and a clock period of *NT*. *N* is assumed to be a power of two. This is a simple circuit to build using D-type flip-flops.

A basic schematic for a divide-by-2 circuit is shown in Figure 16.28. This uses a DFF with a feedback loop to connect $\bar{q}$ back to the input *d* on every rising edge. The divide-by-2 output *Out_2* is taken from the *q* terminal. To understand the operation, note that *q* is assigned the value of *D* at the rising clock edge and remains at that value for one clock cycle. The feedback loop connects $\bar{q}$ back to *d*, so it takes two clock cycles for *Out_2* to complete a sequence of 0 to 1, then back to 0 again. A divide-by-4 counter can be built by adding another stage that uses *Out_2* as the clock. This is shown in Figure 16.29. Additional stages can be added to obtain divide-by-8, divide-by-16, and so on.

Both circuits are easily constructed in Dsch using the DFF element from the **Symbol** library. The only difference is that the library entry is a negative edge-triggered device, but that will not make any difference in the operation. Also note that the DFF must be pro-

Figure 16.28:
Divide-by-2 circuit

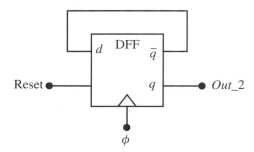

Figure 16.29:
Divide-by-4
network

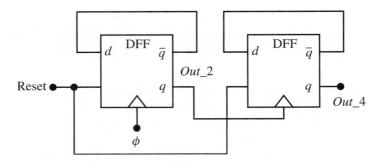

vided with a reset pulse to initiate the operation. The timing diagrams for the divide-by-4 circuit are shown in Figure 16.30; both *Out_2* and *Out_4* are seen to have the expected behavior. The Verilog module listing is obtained in the form:

```
module A_ClockDiv4(Clock,Reset,ClockDiv2,ClockDiv4);
    input Clock,Reset;
    output ClockDiv2,ClockDiv4;
    dreg #(19) dreg1(ClockDiv4,w4,w4,Reset,ClockDiv2);
    dreg #(19) dreg3(ClockDiv2,w6,w6,Reset,Clock);
endmodule
```

The DFFs are called **dregs**, and are predefined Microwind macros.

Microwind produces the compiled divide-by-2 layout shown in Figure 16.31. The divide-by-4 circuit is shown in Figure 16.32 for comparison. The two-cell cascade is evident in the pattern. Larger values of N can be obtained by continuing to chain-on additional cells. The process can be done by creating a symbol for a divide-by-2 network, then duplicating it as needed. The symbolic design of a divide-by-16 circuit is shown in Figure 16.33. This can be used to compile the CMOS network in the same manner.

Figure 16.30:
Timing diagrams
for the divide-by-4
circuit

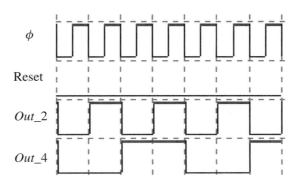

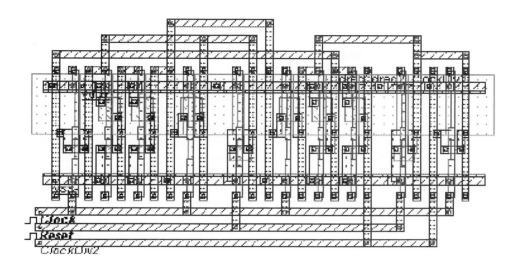

Figure 16.31:   Compiled divide-by-2 circuit

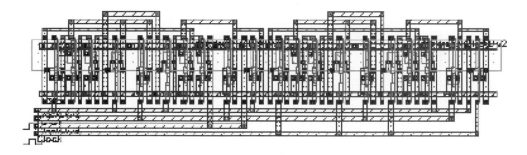

Figure 16.32:   Divide-by-4 circuit

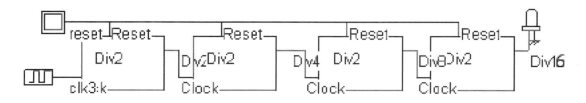

Figure 16.33:   Divide-by-16 circuit

● ● ● ● ● ● ● ● ● ● ● ● ● ● ● ⋅ ⋅ ⋅

16.5 Binary Counter

A binary counter is an important unit in system design. In this example, we will investigate the design of a 4-bit up–down counter using edge-triggered DFFs.

The entire circuit is shown in Figure 16.34. The circuit counts clock pulses and displays the results in the 4-bit word $q_3q_2q_1q_0$. The upward or downward behavior is determined by the *Up/Down* control bit; a value of *Up/Down* = 0 yields an upward counter, while *Up/Down* = 1 gives a downward count. The schematic can be translated directly into a Verilog file and compiled by Microwind to produce the CMOS circuit. This results in a single, horizontal layout. However, note that a primitive cell can be identified, as shown by the dashed box. Compiling this circuit and then replicating it allows us to create different layout geometries.

The structure of one cell is shown in Figure 16.35. To create the 4-bit counter, we duplicate this cell three times. One arrangement is the vertical stacking shown in Figure 16.36; adding the interconnect wiring between the cells completes the design. A 2×2 cell array is another possibility. This demonstrates how a basic cell can be arranged physically on the silicon to produce a logical unit with the desired shape.

Figure 16.34:
Logic circuit for a 4-bit up–down counter

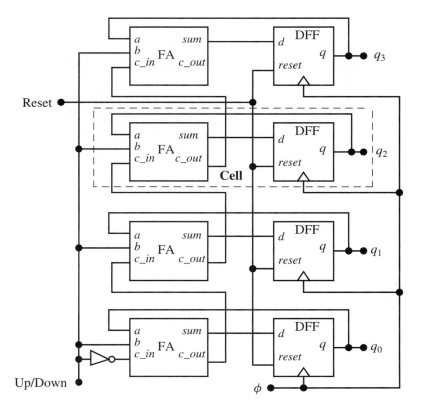

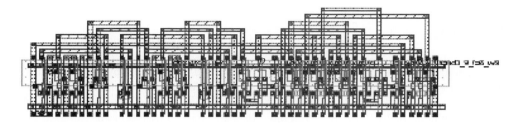

Figure 16.35:   Compiled CMOS circuit for one cell

Figure 16.36:
Array for the 4-bit
counter (unwired)

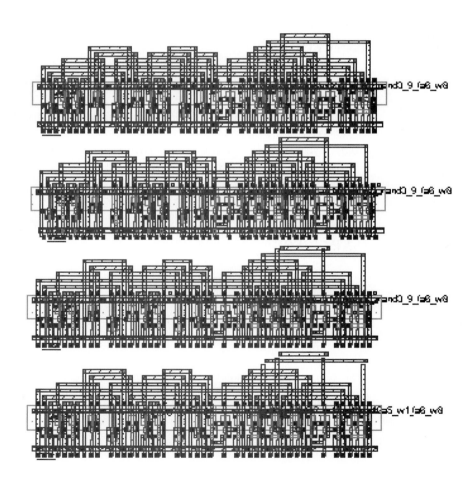

● ● ● ● ● ● ● ● ● ● ● ● ● ● ● ● ●

16.6 Summary

Automated design is a powerful technique that can be used to create complex logic circuits very quickly. The results vary with the complier and the manner in which the flow is applied.

The evolution of VLSI circuits with more than 100-million transistors has spurred on the development of CAD tools that provide increased levels of sophistication. The savvy chip designer is aware of the toolsets and what they can accomplish, but also understands their limits and trade-offs.

● ● ● ● ● ● ● ● ● ● ● ● ● ● ● ● ●

16.7 References

[16.1] Ciletti, M. D., *Advanced Digital Design with the Verilog HDL*. Upper Saddle River, NJ: Prentice Hall, 2003.

[16.2] Smith, M. J. S., *Application-Specific Integrated Circuits*. Reading, MA: Addison-Wesley, 1997.

[16.3] Uyemura, J. P., *Introduction to VLSI Circuits and Systems*. New York: John Wiley & Sons, 2002.

● ● ● ● ● ● ● ● ● ● ● ● ● ● ● ● ●

16.8 Exercises

16.1 Construct a schematic for a 2:1 MUX.

16.2 Use the 2:1 MUX to build a 4:1 MUX.

16.3 Use the 4:1 MUX to build an 8:1 MUX. Export to Verilog and compile the design in Microwind.

Capacitors and Inductors
On-Chip Passive Elements

Digital CMOS switching circuits consist only of MOSFETs wired together with metal interconnects. An analog circuit, on the other hand, may need a capacitor or an inductor to achieve the desired functions. While capacitors have usually been fabricated on the chip itself, the introduction of integrated inductors is a more recent development.

This short chapter introduces these on-chip passive elements and presents the technique of generating an inductor layout geometry in Microwind. Application details can be found in the referenced texts at the end of the chapter.

● ● ● ● ● ● ● ● ● ● ● ● ● ● ● ● · · ·

17.1 Integrated Capacitors

Every node in a CMOS circuit is intrinsically capacitive, but a typical value is only a few femtofarads (fF). Larger capacitors are used for frequency compensation, filter design, and other analog-type functions. The chip designer cannot depend upon using parasitics for these elements, because the value of the capacitor is critical to the operation.

17.1.1 Poly–Poly Capacitors

A classical approach to building on-chip capacitors in a CMOS process flow is to use a second layer of polysilicon denoted as Poly2. The first layer of polysilicon (Poly) is used for FET gates, the Poly layer is coated with a thin insulating oxide layer, and then Poly2 is deposited on the wafer. This gives a Poly–Poly2 capacitor structure, as shown in Figure 17.1(a). The insulator thickness, t_{ins}, determines the capacitance per unit area. The top view in Figure 17.1(b) defines the capacitor area as the Poly2 area, A_2, that is projected down onto the bottom Poly plate. The parallel-plate capacitor formula gives the capacitance:

Figure 17.1:
Poly–Poly2
capacitor

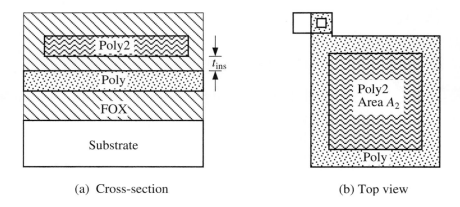

(a) Cross-section

(b) Top view

$$C_{P-P2} = \frac{\varepsilon_{ins}}{t_{ins}} A_2 \tag{17.1}$$

by ignoring fringing effects. The Poly–Metal1 contact is shown explicitly in the top-view drawing. Access to the top Poly2 plate is achieved by adding a Metal–Poly2 contact.

Although Poly–Poly capacitors have been very popular in analog processes, only a few processes that have the second polysilicon layer need to use this structure. A recent survey of the technologies available through MOSIS listed only two active fabrication lines that offered Poly–Poly capacitors. The smallest listed process was 0.35 μm. This probably is due to the fact that the newer fabrication technologies are oriented to digital circuits that do not require a second poly layer. When they are needed, metal layers are used instead. These are simpler and are placed away from the silicon surface area.

17.1.2 Metal–Metal Capacitors

This type of structure is easily built in multiple-metal layer processes. The capacitors are formed using patterns on two metal layers, Metal$x$ and Metal$y$, such that:

$$C_{MM} = C_{xy} A_{ov} \tag{17.2}$$

where C_{xy} is the capacitance per unit area and A_{ov} is the overlap area. The insulator thickness may be the standard value, but thin oxide options may be offered for larger values of C_{xy}.

Capacitor plates can be quite large for even small values of C. Assuming a square geometry with dimensions of $d \times d$, the required side length is given by

$$d = \sqrt{\frac{C}{C_{xy}}} \tag{17.3}$$

As an example, the default value of C_{xy} between Metal4 and Metal5 layers in the Microwind 0.12 μm process is 100 aF/μm^2. A capacitor with a value of 5 fF = 5,000,000 aF has a side length of

$$d = \sqrt{\frac{5,000,000}{100}} \approx 223 \quad \mu\text{m} \tag{17.4}$$

and a total area of 50,000 μm^2. The gate area of a minimum-sized FET is about

$$0.12 \ \mu\text{m} \times 0.24 \ \mu\text{m} = 0.0288 \ \mu\text{m}^2$$

so that the 5 pF capacitor is 50,000/0.0288 or 1,736,111 times larger! If a thin insulator is used instead, then the area of the capacitor can be decreased accordingly. However, capacitors are still considered "area hogs" and are used sparingly in design.

17.1.3 MOSFET Capacitors

The gate capacitance of a MOSFET also can be used to create a capacitor by shorting both the drain and the source connections. Figure 17.2 illustrates the wiring and a common symbol for this type of element.

Denoting the area of the gate as $A_G = W_L$, the capacitance is calculated from

$$C = \frac{\varepsilon_{\text{ox}}}{t_{\text{ox}}} A_G = C_{\text{ox}} WL \tag{17.5}$$

Since the thickness of the gate oxide, t_{ox}, is very small (less than about 3 nm), the oxide capacitance per unit area, C_{ox}, is larger than other layer-to-layer values on the chip. To obtain the full value of C, the gate voltage should be greater than V_{Tn} and positive with respect to the drain/source terminal.[1] Although the simplest conceptual design for this type of capacitor is a square of side length, d, many processing lines do not permit arbitrary values for the FET gate length, L. In this case, a standard FET layout strategy must be used.

Figure 17.2:
MOSFET used as a capacitor

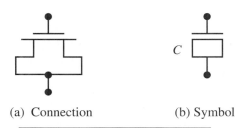

(a) Connection   (b) Symbol

1. This insures that the electron-channel layer is formed underneath the oxide.

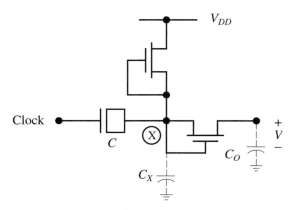

An example of a FET capacitor circuit is the charge pump in Figure 17.3. This takes a full-rail input clock and dynamically boosts the voltage at node X to a maximum value of

$$V_X = (V_{DD} - V_{Tn}) + \frac{C}{C + C_X} V_{DD} \qquad (17.6)$$

over a number of clock cycles. In this expression, C_X is the total parasitic capacitance at the node. If $C \gg C_X$, then V_X can be boosted to a value

$$V_X \approx 2 V_{DD} - V_{Tn} \qquad (17.7)$$

The output voltage then reaches a maximum value of $V \approx 2(V_{DD} - V_{Tn})$. Charge pumps are useful for boosting voltages and for the local biasing of wells.

17.1.4   Common Centroid Layout

Analog circuits are quite sensitive to variations in the processing. At the layout level, point-to-point variations can cause problems when balanced element values are required.

One problem that arises in certain types of **Digital–Analog Converters (DACs)** is the need for a set of capacitors with values that are related by a scaling factor. For example, a charge-scaling DAC may use capacitors with values C, $2C$, and $4C$; and the precision of the converter depends upon having these ratios. Baker [17.2] The required plate areas are A, $2A$, and $4A$; but the effects of process variations will be different because of the differences in the edge periphery effects, such as fringing and non-uniform etching. A solution is to use $(1 + 2 + 4) = 7$ plates with area A and wire them as necessary to form the three capacitors. A **common centroid** arrangement is shown in Figure 17.4. The single C element (7) is placed in the center. The $2C$ capacitor is obtained by wiring the two central elements (2 and 5) together, while the $4C$ capacitor is obtained by wiring each of the remaining elements (1, 3, 4, and 6) in parallel. This distributes the base components out in a balanced, geometrical manner around a common center. Common centroid layouts are

Figure 17.4:
Common-centroid
capacitor layout

Figure 17.4:
Common-centroid
capacitor layout

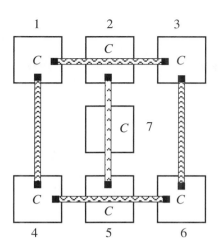

used to combat process variations, such as changes in the insulator thickness across the region. This technique generally gives better ratioing characteristics among the element values.

Common-centroid layout is used when it is important to obtain element values with specified ratios; the exact values are demoted to secondary importance. For example, a circuit may need a set of capacitors where $C_1 = C$, $C_2 = 4C$, and $C_3 = 8C$; but the numerical value of C is not important so long as it is within a specified range.

Consider the situation illustrated in Figure 17.5. Two elements, A and B, are to be matched so they have the same layout geometry. However, processing variations can cause the two to have different electrical characteristics. For example, if the process parameters are constant in the x-direction but have variations in the y-direction, the two will not be identical. If the gradients are along arbitrary directions, such as along the dashed line shown, then the matching problem becomes very complicated. The common-centroid approach attacks this problem by breaking the device into segments around a center point and then using symmetrical wiring to obtain the final components.

Three 2-axis common-centroid solutions are shown in Figure 17.6. Each group has a center point around which the array is weighted. Several other possibilities exist for both 1- and 2-axis patterns. Some single-axis, 2-element patterns are shown in Figure 17.7. The technique can be expanded to include three or more elements in the same manner. While the arrangements seem straightforward, wiring issues can complicate the designs if the

Figure 17.5:
Directional process
variations

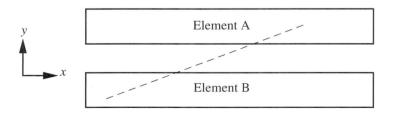

Figure 17.6:
Dual-axis common-centroid examples

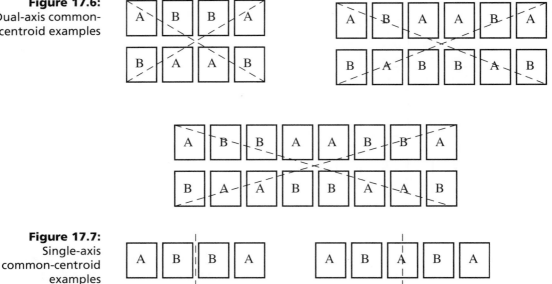

Figure 17.7:
Single-axis
common-centroid
examples

number of elements get too large. Dual-axis arrays can be particularly difficult to wire with the desired symmetry. Hastings [17.3] has an extensive list of examples and an excellent discussion of the main ideas.

The idea of common-centroid layout can also be applied to multiple-gate finger MOSFETs that have large W values. This is discussed in more detail in the next chapter.

● ● ● ● ● ● ● ● ● ● ● ● ● ● ● ● ● ●

17.2   Integrated Inductors

High-frequency circuits have worked their way down to the chip level. Integrated CMOS inductors are becoming more common in Radio Frequency (RF) chips that are used for telecommunications and wireless applications. An on-chip inductor increases the reliability over that of an off-chip design, since it eliminates the external wiring failures. However, inductors are difficult to design and can create other problems in the on-chip environment.

Inductors store magnetic energy. The most common macro-design is a simple coil of insulated wire that carries a current, i, which generates a magnetic field, H, as described by Faraday's law. The magnetic-field lines cut closed areas of the coil, giving rise to a **magnetic flux**, Φ. The inductance is the proportionality constant between Φ and i as seen in

$$\Phi = Li \tag{17.8}$$

The current–voltage relationship is based on **Faraday's law**, such that

$$v = \frac{d\Phi}{dt} = L\,\frac{di}{dt} \tag{17.9}$$

This excursion into basic physics shows that inductors intrinsically are associated with currents and magnetic fields. It is not possible to shield magnetic fields with any material layers on a CMOS chip, so we must be careful about the placement of an integrated inductor to insure that the magnetic field lines do not cause problems in the rest of the circuit. Controlling magnetic crosstalk and associated phenomena are critical to the functionality of a high-frequency circuit.

Integrated CMOS inductors did not become practical until the processing technology advanced to the point of providing three or more layers of metal. The top-most metal layer is the most common choice for an inductor, as it raises the element above the silicon circuitry with the largest possible distance of separation. Moreover, several processes allow the chip designer to specify a **thick metal** option for the top layer, which decreases the sheet resistance. This results in a higher **resonance quality**, **Q**, as a factor for the device.

The simplest geometry for an integrated inductor is the single-layer spiral. In a Manhattan geometry, the top view of the structure appears as seen in Figure 17.8. The defining parameters are

H = overall height/width

w = conductor width

s = conductor spacing

Also, we define

n = number of turns

The inductance, L, increases with the number of turns, and even nanoHenry (nH) inductors can require large areas.

Figure 17.8:
Basic inductor layout

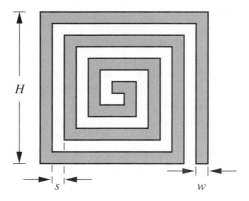

Although it may seem that finding the value of L is an easy task, inductances are never simple to calculate. Inductances of structures can be estimated using empirical formulas, but numerical calculations of the full set of four Maxwell equations usually are required for accurate values. A simple approximation for the inductance is

$$L \approx 1.2 \times 10^{-6} \left(\frac{n^2 H}{2} \right) \tag{17.10}$$

which models the structure as a simple spiral. Microwind estimates the inductance of the spiral structure using Lee[17.4]

$$L \approx 37.5 \mu_o \left(\frac{n^2 a^2}{22H - 14a} \right) \text{ in } \mu H \tag{17.11}$$

where with n, a, and H are in μm, and $\mu_o = 4\pi \times 10^{-7}$ H/m is the free-space permeability, and a is the mean radius of the spiral as measured from the center to the middle of the windings. Razavi [17.5] suggests using

$$L \approx 1.3 \times 10^{-7} \frac{A_m^{5/3}}{A^{1/6} w^{7/8} (w+s)^{1/4}} \text{ in } \mu H \tag{17.12}$$

where A_m, A are in μm^2, and A_m is the area of the metal and A is the total area of the inductor. The units for ω and s are in μm.

Microwind has an inductor layout as a standard feature. To access the **Layout Generator**, execute the Menu command sequence:

Edit $\Rightarrow$

    Generate $\Rightarrow$

        Inductor

This opens the Inductor **Layout Generator** window shown in Figure 17.9. This allows you to specify the number of turns, the width, and the spacing of the conductor lines. The inductor will be placed on the metal layer selected on the right side of the screen. To create an inductor, fill in the fields with the desired values. Pushing the **Update L, Q** button provides an estimate of the inductor if it were a straight line; this is usually much larger than the value for spiral geometry. Use the **Generate Inductor** button to place the layout into the work area. You should execute a few **Zoom Out** commands to accommodate the large size; you can always adjust the final view later.

An example of a Microwind-generated layout is shown in Figure 17.10. A lambda ruler has been added to illustrate the large size of the element compared with a basic circuit. The **Navigator** may be used to view the electrical characteristics. This provides the basic capacitance, resistance, and inductance of any node. The screen in Figure 17.11

Figure 17.9:
Inductor Layout
Generator window

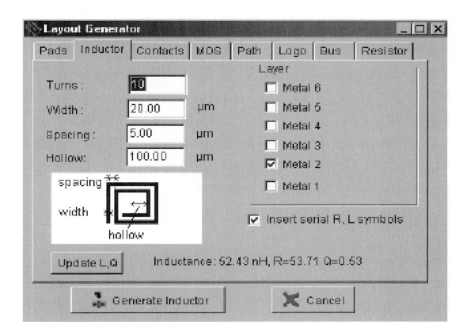

Figure 17.10:
Inductor layout
example

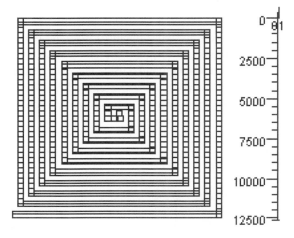

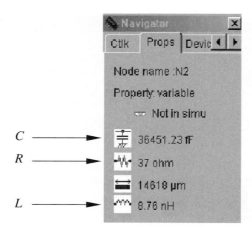

shows typical values; it should be noted that the large area of the conductor trace leads to a high capacitance with respect to the substrate. Microwind computes L using Equation (17.11). It also estimates the quality factor, Q, using

$$Q = \frac{1}{R}\sqrt{\frac{L}{C_1 + C_2}} \qquad (17.13)$$

with the model shown in Figure 17.12.

The reader is cautioned not to interpret the Microwind values too literally, as they are only *estimates*. The inductance and Q-value of a metal spiral pattern varies with the thickness of the layer and is also affected by other metal structures that couple to it (i.e., interconnect). The best estimates are those obtained from the foundry tests and design sheets for a specific process. Empirical formulas are also useful for obtaining first design estimates. The actual values can be computed using electromagnetic equation solvers.

Another problem is that an inductor can create eddy currents in the substrate, and this effect is enhanced with heavy doping. The Q is lowered because of the resistance, and the

Figure 17.12:
Model used to
compute the
quality factor, Q

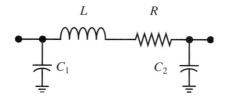

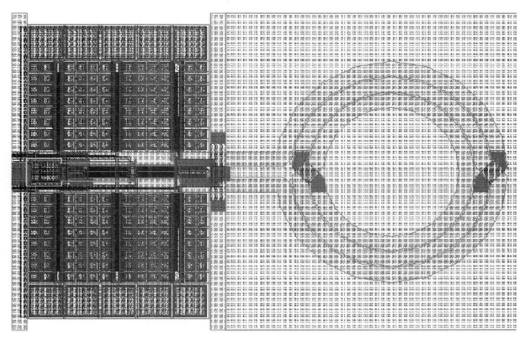

Figure 17.13:   Inductor and circuit layout example

currents themselves can interfere with circuit operation. Substrate regions below inductors are usually devoid of transistor circuitry for this reason. The design rules will stipulate the region affected, and they should be followed to ensure proper operation.

Inductors can also be designed manually. An example of an **inductor–capacitor (LC)** oscillator coil designed at Georgia Tech is shown in the layout of Figure 17.13. The circular geometry was created using small squares on the outer edges, and the inductor was created on the highest metal layer in the CMOS process. The square sections to the left are the transistors and tuning capacitors. Figure 17.14 shows a photograph of the fabricated structure in a 0.18 μm technology. The design is quite different than the basic square geometry but has similar features.[2]

The use of integrated inductors in CMOS RF circuits is a relatively recent development. With the importance of high-frequency electronics in communications, knowledge and techniques will grow quickly to the point where inductors can be included easily in a chip design.

---

2.   This particular structure was designed by Dr. Paul Murtagh of Integrated Device Technology (IDT) in conjunction with Dr. Declan McDonagh, Dr. Brian Butka, and Dr. Y. Alper Eken.

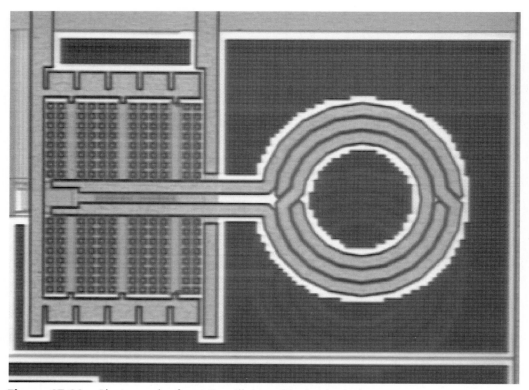

Figure 17.14: Photograph of an LC oscillator with circular inductor

17.3 References

[17.1] Allen, P. E., and Holburg, D., *Analog CMOS*. Oxford University Press, 2002.

[17.2] Baker, R. J., Li, H. W., and Boyce, D. E., *CMOS Circuit Design, Layout, and Simulation*. Piscataway, NJ: IEEE Press, 1998.

[17.3] Hastings, A., *The Art of Analog Layout*. Upper Saddle River, NJ: Prentice Hall, 2001.

[17.4] Lee, T. H., *The Design of CMOS Radio-Frequency Integrated Circuits*. Cambridge, U. K.: Cambridge University Press, 1998.

[17.5] Razavi, B., *RF Microelectronics*. Upper Saddle River, NJ: Prentice Hall, 1998.

● ● ● ● ● ● ● ● ● ● ● ● ● ● ● ● ● ● ●

17.4   **Exercises**

17.1   Design a FET capacitor with a value of 20 fF in your choice of technologies.

17.2   Use the Microwind Inductor Layout Generator to create a coil with four turns and default line width and spacing values.

(a)   Use the Navigator to find the estimated inductance.

(b)   Calculate the inductance as predicted by Equation (17.11).

(c)   Calculate the inductance using Equation (17.12).

Analog CMOS Circuits Layout Basics

Analog CMOS design techniques can be quite different from those used in digital systems. Instead of worrying about switching times, concern is directed towards frequency response, biasing, and linearity. Component matching becomes important, and parasitics enter the problem in a very different way. This chapter is a brief introduction to some common analog CMOS circuits in silicon. Several excellent books are listed at the end of the chapter for those desiring a more in-depth discussion.

18.1 Simple Amplifiers

The amplifier is one of the most basic components in analog design. It provides gain and is the basis of many other circuits. In this section, we examine the gain characteristics of basic circuits. The equations provide insight into some of the layout features that need to be considered.

A basic nFET amplifier is shown in Figure 18.1. This circuit uses a current source, I_1, to bias the transistor. It is assumed that M1 is operating in saturation. The important concepts can be developed using the simple Square Law equation:

$$I_D = \frac{1}{2}k_n'\left(\frac{W}{L}\right)_1 (V_{GS} - V_{Tn})^2[1 + \lambda V_{DS}] \tag{18.1}$$

to describe the transistor. The small-signal transconductance of the nFET is given by

$$g_m = \left.\frac{\partial I_D}{\partial V_{GS}}\right|_{V_{DS}= \text{constant}} \tag{18.2}$$

Figure 18.1:
Basic nFET amplifier

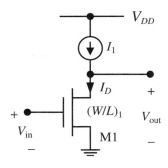

which has units of A/V = S. Differentiation of the current yields

$$g_m = k_n' \left(\frac{W}{L}\right)_1 (V_{GS} - V_{Tn})[1 + \lambda V_{DS}] \tag{18.3}$$

This can be cast into the alternate form:

$$g_m = \sqrt{2k_n' \left(\frac{W}{L}\right)_1 I_D(1 + \lambda V_{DS})} \tag{18.4}$$

Both show that g_m is proportional to the **aspect ratio**, $(W/L)_1$.

The small-signal voltage gain of the amplifier is

$$A_v = \frac{v_{out}}{v_{in}} = -g_m r_o \tag{18.5}$$

where

$$r_o = \frac{\partial V_{DS}}{\partial I_D} \tag{18.6}$$

is the **small-signal output resistance** of the FET, and it has been assumed that the current source has an infinite input impedance. This can be written in the form:

$$r_o = \frac{1}{\frac{1}{2}k_n' \left(\frac{W}{L}\right)_1 (V_{GS} - V_{Tn})^2 \lambda} \approx \frac{1}{\lambda I_D} \tag{18.7}$$

With an external load resistance, R_L, the gain equation becomes

$$A_v = -g_m(r_o || R_L) \tag{18.8}$$

Figure 18.2:
Diode-connected
nFET

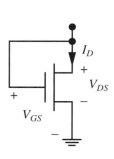

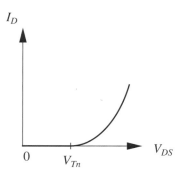

If $r_o \gg R_L$, then

$$A_v \approx -g_m R_L \tag{18.9}$$

The gain is proportional to g_m, which is itself proportional to $(W/L)_1$. This implies that large FETs are used to increase the gain of the stage. The f_T of an nFET is given by

$$f_T = \frac{g_m}{(C_{GS} + C_{GD})} \tag{18.10}$$

which remains approximately constant with increasing size.

Another characteristic of analog design can be obtained by replacing the ideal current source with the **diode-connected** FET shown in Figure 18.2(a). This is just a transistor with the drain and the gate wired together. The device isn't really a diode, but gains its name from the observation that the I–V characteristics resemble that of a pn-junction diode.[1] To understand this comment, note that $V_{DS} = V_{GS}$. If $V_{GS} \leq V_{Tn}$, the transistor is in cutoff with

$$I_D \approx 0 \tag{18.11}$$

For $V_{GS} > V_{Tn}$, the transistor is saturated with

$$I_D = \frac{1}{2}k_n' \left(\frac{W}{L}\right)_1 (V_{DS} - V_{Tn})^2 [1 + \lambda V_{DS}] \tag{18.12}$$

This gives the plot shown in Figure 18.2(b). This has a turn-on characteristic at the threshold voltage, V_{Tn}, which is similar to the starting conduction through a diode. It is noted in passing that diode-connected FETs were used in the charge-pump example of the previous chapter.

---

1. Also, the circuit is analogous to an npn bipolar transistor with its collector and base connected together, which forms a physical pn-junction device.

Figure 18.3:
Amplifiers with a
diode-connected
FET load

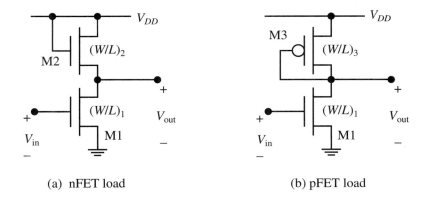

(a) nFET load   (b) pFET load

The amplifier in Figure 18.3(a) uses a diode-connected nFET as a load. Analyzing the small-signal voltage gain for this circuit gives

$$A_v = -g_{m1}\left(\frac{1}{g_{m2} + g_{mb2}}\right) \qquad (18.13)$$

where

$$g_{mb2} = \frac{\partial I_{D2}}{\partial V_{BS2}} = \eta g_{m2} \qquad (18.14)$$

is due to body-bias effects in M2 through the threshold-voltage expression

$$V_{Tn} = V_{Tn0} + \gamma(\sqrt{2|\phi_F| + V_{SB}} - \sqrt{2|\phi_F|}) \qquad (18.15)$$

such that

$$\eta = \frac{\gamma}{2\sqrt{2|\phi_F| + V_{SB}}} \qquad (18.16)$$

The gain can be rewritten in the form

$$A_v = -\sqrt{\frac{(W/L)_1}{(W/L)_2}}\left(\frac{1}{1 + \eta}\right) \qquad (18.17)$$

which demonstrates another important idea in analog design:

The performance of a circuit may depend upon a ratio of device sizes.

In this case, the gain depends only on the relative size of M1 to M2, and not on the individual aspect ratios. We note in passing that some digital CMOS circuits, such as pseudo-nMOS gates, have the same characteristic.

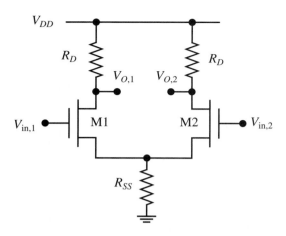

A diode-connected pFET is used as a load in Figure 18.3(b). The pFET is wired such that $V_{SG} = V_{SD}$ and conducts current if $V_{SG} > |V_{Tp}|$ but is in cutoff for $V_{SG} \leq |V_{Tp}|$. The gain for the amplifier circuit is

$$A_v = -\sqrt{\frac{\mu_n (W/L)_1}{\mu_p (W/L)_3}}$$

(18.18)

where the mobility factors distinguish between the conduction mechanisms of the two FET polarities. In this case, the pFET bulk and the source are both at V_{DD}, so there are no body-bias effects. However, the dependence on the device sizes is still apparent.

Another example is the basic differential amplifier in Figure 18.4. This is designed as a balanced circuit with identical transistors and load resistors, R_D. The small-signal differential gain of the circuit is given by

$$A_{dm} = -g_m (r_o || R_D)$$

(18.19)

while the common-mode gain is

$$A_{cm} = -\frac{g_m (r_o || R_D)}{1 + 2g_m R_{SS}}$$

(18.20)

The **common-mode rejection ratio (CMRR)** for a balanced output is

$$CMRR = 1 + 2g_m R_{SS}$$

(18.21)

These equations assume that the left and right sides are indeed identical.

Figure 18.5:
FET current mirror

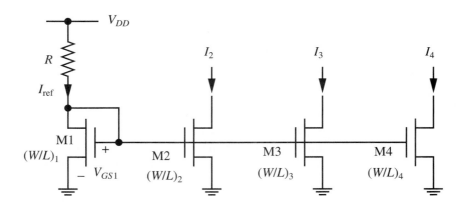

Device mismatches complicate the situation. Suppose that the operating points of M1 and M2 are not the same, resulting in distinct values for g_{m1} and g_{m2}. This could occur if M1 and M2 are not matched or if the resistors are not equal. In this case, the *CMRR* is changed to

$$CMRR = \frac{g_m}{(g_m \pm (\Delta g_m/2))} + 2g_m R_{SS} \tag{18.22}$$

where

$$g_m = \frac{1}{2}(g_{m1} + g_{m2}) \tag{18.23}$$

is the simple average and

$$\Delta g_m = g_{m1} - g_{m2} \tag{18.24}$$

is the difference. Obviously, a large value of Δg_m reduces the CMRR and leads to inferior rejection characteristics.

As a final example, consider the simple MOS current mirror shown in Figure 18.5. This uses the current,

$$I_{\text{ref}} = \frac{V_{DD} - V_{GS1}}{R} \tag{18.25}$$

as a reference value. Assuming that M1 is saturated,

$$I_{\text{ref}} = \frac{1}{2}k_n' \left(\frac{W}{L}\right)_1 (V_{GS1} - V_{Tn})^2 [1 + \lambda V_{DS1}] \tag{18.26}$$

The gate terminal is common to the other transistors, so that every FET has the same value of $V_{GS} = V_{GS1}$. The current, I_N, for $N = 2, 3, 4$ is given by

$$I_N = \frac{(W/L)_N}{(W/L)_1}\left(\frac{1 + \lambda V_{DSN}}{1 + \lambda V_{DS1}}\right)I_{\text{ref}} \qquad (18.27)$$

It is obvious that mismatches in the aspect ratios will affect the currents. Processing variations are also important, as the threshold voltages may be slightly different.

The examples in this section have been chosen to illustrate that analog circuits tend to use larger transistors than those employed in digital design. Moreover, device matching can be critical to the operation, so that analog layout is more sensitive to layout-induced mismatching. The use of symmetries and common centroids becomes mandatory.

18.2 MOSFETs

The layout procedures for analog FETs are the same as for digital designs. However, the discussion of the previous section indicates that the problems of uniformity and balancing are more critical because of the sensitivity of analog circuits to variations in electrical parameters.

18.2.1 Large Transistors

MOSFETs with large **channel widths**, W, usually employ multiple gate fingers. One reason is to obtain an overall shape that is easier to integrate into the layout. Long, skinny transistors are difficult to place and wire, while FETs that have a shape closer to a square are much easier to deal with. Another reason for using multiple gates is because the resistance of a long, poly gate can get large.

An example is shown in Figure 18.6(a), where a long poly, gate has an input resistance of

$$R_g = R_{s,\,\text{poly}}\left(\frac{W}{L}\right) \qquad (18.28)$$

for a current that flows into the gate electrode to charge the gate capacitance. The **poly sheet resistance**, $R_{s,\text{poly}}$, can be large, even in a salicided device.

Figure 18.6:
Single and
multiple-gate
finger layouts

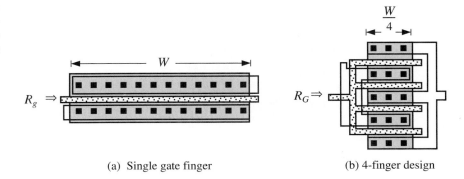

(a) Single gate finger   (b) 4-finger design

The layout for a 4-fingered design is shown in Figure 18.6(b). This has four transistors that are wired in parallel. Each embedded FET has a channel width of $W/4$. This results in a shape that is much more compact in the horizontal direction. The poly resistance of a finger is

$$R_{g4} = \frac{1}{4}R_g \tag{18.29}$$

and the fingers are in parallel. A typical rule of thumb in analog design is to select the number of fingers so that the resistance of a finger is less than $1/g_m$ of the transistor associated with the finger. We also note that the drain–source parasitic resistances are in parallel, so they will also be reduced. These techniques also can be applied to digital circuits to reduce the gate time constant.

Multiple-gate finger designs can also reduce the node capacitances. The left-metal terminal only sees two doped regions, while the right-metal node is connected to three. Since the width has been reduced to $W/4$, the parasitic junction capacitances may be smaller than in the long design. The actual values are based on the relative junction areas.

In critical applications, even small variations of the transistor characteristics may be important. In Figure 18.6(b), the two center-gate fingers have different surroundings than the two outer fingers. Nonuniform etching or doping may result in a situation where the outer fingers have slightly different operational characteristics, so that the overall structure does not behave as intended. Dummy elements can be used to overcome this type of a problem. Figure 18.7 shows a 4-fingered layout in which dummy poly gate lines have been added on both sides of the device. Using this ensures that every active gate finger sees the same surroundings on both the left and right sides. Also, this transfers any edge-specific processing variations to the dummy gate lines. Note that the entire structure is balanced using the concept of a 1-dimensional common centroid.

Figure 18.7:
Use of dummy lines
for balanced layout

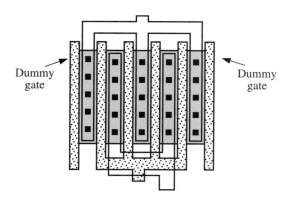

Dummy
gate

Dummy
gate

18.2.2 Matching Transistors

Circuits, such as differential amplifiers and current mirrors, are sensitive to differences in the amplifying devices. Processing variations can be random in nature, making it difficult to match two (or more) transistors. First, let us adopt the analog nFET and pFET symbols shown in Figure 18.8. The arrow is used to denote the direction of current flow and is placed at the source. Although simple FETs are symmetric, this may not be the case in large devices, so this notation distinguishes between the sides.

Consider the problem of matching two amplifier FETs for a differential amplifier such as in Figure 18.9. For balanced operation, nFETs, Mn1 and Mn2 should be identical, as should Mp1 and Mp2. With a balanced circuit, the differential mode gain is given by

$$A_v = -\sqrt{\frac{\mu_n (W/L)_n}{\mu_p (W/L)_p}} \qquad (18.30)$$

but the *CMRR* is reduced if the circuit is not balanced.

A straightforward technique for balanced FET layout is shown in Figure 18.10. Single-finger devices are shown in Figure 18.10(a), while larger transistors are created using the dual-gate FETs in Figure 18.10(b). These use a single-axis common-centroid idea to balanced the transistors around the I_{SS} line. Although nFETs have been used, the idea applies equally well to pFETs. Note that the higher-capacitance side has been connected to

Figure 18.8:
FET symbols used in
analog circuits

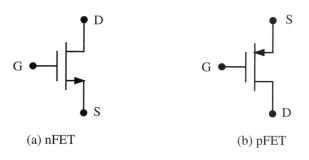

(a) nFET

(b) pFET

Figure 18.9:
Differential
amplifier with
active loads

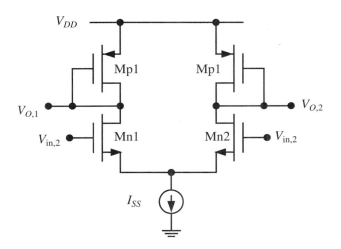

I_{SS} in Figure 18.10(b); this reduces the capacitance on output nodes, improving the frequency response of the amplifier.

The dual-transistor layout can be transformed into a single, large-width FET, as shown in Figure 18.11(a). This design is slightly asymmetric in the vertical direction because of the metal wiring. Using a perpendicular Metal2 line and adjusting the location of the gate ports, result in the more balanced structure in Figure 18.11(b).

Although it is possible to design individual transistors and matched groups using techniques like these, the wiring gets complex. Figure 18.12 shows a pair of source-coupled nFETs that have been designed using four fingers for each device in a dual-axis common-centroid arrangement. The gate connects have been omitted for clarity. The sequence is seen to be

M1-M2-M2-M1

M2-M1-M1-M2

Figure 18.10:
Balanced transistor
layout

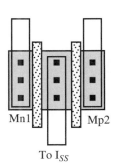

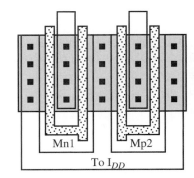

(a) Single-fingered design

(b) Dual-fingered transistors

Figure 18.11:
Balanced
large-width
transistor

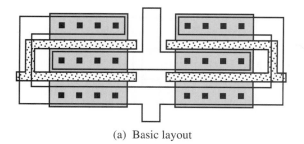

(a) Basic layout

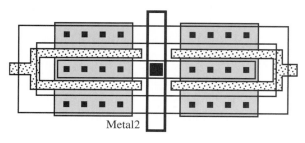

(b) Balanced wiring

and the terminals can be identified by tracing each transistor. Increasing the number of fingers makes the wiring much more difficult, but patterns like this can be designed if needed.

Figure 18.12:
Source-coupled
pair with a
common-centroid
design

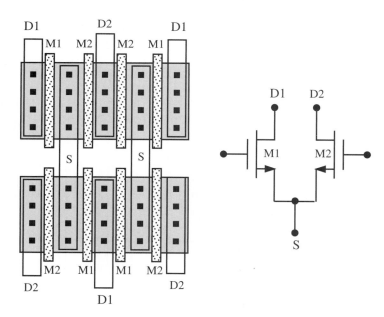

• • • • • • • • • • • • • • • •

18.3 Resistors

Processing variations prohibit the fabrication of individual resistors with precise, predetermined values. This was recognized in the early days of bipolar design and originates from the sensitivity of the resistance formula:

$$R = R_s \frac{l}{w} \tag{18.31}$$

to variations in the sheet resistance and the width. The sheet resistance, R_s, depends on the thickness, t, of the layer by

$$R_s = \frac{\rho}{t} \tag{18.32}$$

so the variation is computed as

$$\Delta R_s = -R_s \left(\frac{\Delta t}{t} \right) \tag{18.33}$$

Since t is small (usually about 0.5 μm or less), even small variations Δt in the thickness can have a large effect on the value of R_s.

It is, however, possible to fabricate a pair of resistors that have approximately the same value of R, even if we cannot predict the value. Owing to this situation, analog integrated circuits that employ resistors use matched pairs as the basis for the operation. This was seen already in the differential amplifier of Figure 18.4. Polysilicon over FOX is probably the most common choice for integrated resistors, since elements on this layer do not consume the silicon area, and unsalicided poly lines can have high sheet-resistance values. If Poly2 is available, it becomes the natural candidate for resistor tracks, since it does not interfere with FETs. Doped active regions (ndiff and pdiff) can also be used, so long as the parasitic junction capacitance and pn junctions are not important.

Designing matched resistors is similar to the problem studied for FETs, except that the length, l, of the structures can be large, which leads to other considerations. First, serpentine patterns are usually required to obtain reasonable resistance values of even a few kilohms.[2] This means that the symmetry is now created with relatively large groups of polygons that follow a zig-zag path.

A basic layout scheme is shown in Figure 18.13. This uses two serpentine patterns in a single-axis balanced arrangement. Although simple, any process variation in the horizon-

---

2. Some submicron processes allow one to specify unsalicided layers, which increases the sheet resistance considerably.

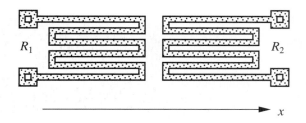

tal (x) direction will create a mismatch with $R_1 \neq R_2$. For example, suppose that the sheet resistance varies locally according to

$$R_s(x) = R_{so}\left(1 - \frac{x}{x_o}\right) \tag{18.34}$$

where R_{so} and x_o are constants. Since most of the resistance lines are horizontal, this decrease in sheet resistance would result in $R_1 > R_2$. Another technique is the interlaced arrangement in Figure 18.14. This uses two series-connected segments to create a resistor. Spreading out the segments by the interlacing helps to equalize the effect of spatial variations.

Some circuits require a set of ratioed resistors. These can be designed using primitive elements that are wired together to form the desired values such that the array provides some balancing. Figure 18.15 shows a group of five resistors with values R_1, $R_2 = 2R_1$ and $R_4 = 4R_1$ in the arrangement R_1 -R_2 -R_4 -R_2 -R_1. This layout equalizes parametric variations in the vertical direction since each segment is affected equally. If we parallel the R_1 elements to create a resistor with a value of $R/2$, and also parallel the R_2 segments to obtain an equivalent resistor with a value of R_1, then the layout becomes balanced (for the new values). Common-centroid techniques can be used to design similar arrays with single- and dual-symmetry axes.

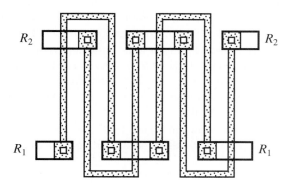

Figure 18.15:
Resistor group
designed from
segments

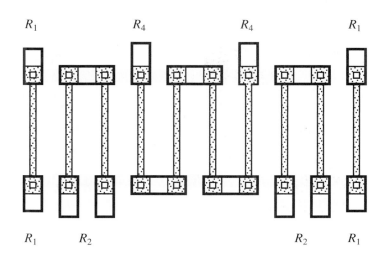

18.4   Signal Wiring

Many analog systems use balanced amplifiers and comparators. The routing of the signal lines becomes important because input and output pairs in these types of circuits are out of phase by 180° and enhanced coupling may occur.

This problem is shown in Figure 18.16. The amplifier outputs, $v_{\text{out},1}$ and $v_{\text{out},2}$, are connected to parallel interconnect wiring, but since they are out of phase, this creates a difference of $(V_1 - V_2)$ in the line voltages. Modeling the coupling capacitance as a lumped element with a value of C_c, results in an imbalance and creates a current flow from line 1 to line 2 of

$$i_{12} = C_c \frac{d}{dt}(V_1 - V_2).$$
(18.35)

Since $V_1 - V_2$ is always changing in time, the crosstalk is enhanced, leading to signal degradation. The same problem occurs in digital logic circuits that employ **dual-rail logic** in

Figure 18.16:
Line coupling with
differential signals

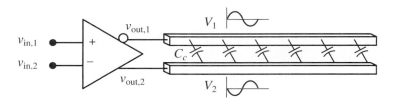

Figure 18.17:
Shielding in parallel
runs

which every variable is accompanied by its complement. A similar problem is **stray coupling** from unrelated lines, which leads to increased noise.

This type of coupling can be minimized by avoiding long runs of parallel signal lines, but that decreases the integration density. A more common solution is to pay careful attention to symmetry and signal routing so that voltages on adjacent lines are in-phase as much as possible. This keeps the difference of $V_1 - V_2$ close to 0 and minimizes the coupling. One approach is illustrated in Figure 18.17. Here, lines that carry similar amplitude in-phase signals (either + or −) are grouped together, and grounded interconnects are used as shields. A random signal $r(t)$ is also shielded using this technique. One difficulty with this technique is that adjacent lines should be *identical* in every respect, and they should experience the same environment along the entire run. This can be very difficult, as illustrated by the simple example in Figure 18.18. Although the two lines follow the same central path, the upper line is longer because of the turns, and the two "see" different surroundings.

An interesting technique for desensitizing a pair of differential signals to the crosstalk induced by a random signal, $r(t)$, is to expose both lines to the same amount of crosstalk by using a crossover. Figure 18.19 shows the main idea. This turns the crosstalk into a common-mode signal, which is then rejected by the amplifier (assuming it has a high *CMRR*, of course).

Figure 18.18:
Turns in parallel
lines

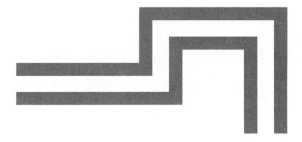

Figure 18.19:
Using
common-mode
rejection to reduce
coupled noise

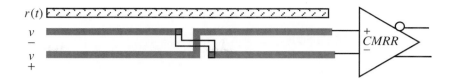

Routing techniques tend to grow more sophisticated as the frequency of the circuits is increased. In multi-GHz CMOS designs, the wiring is sometimes more difficult than designing the components. Design groups that are engaged in this type of design tend to develop a set of rules that help to ensure a functional circuit on the first silicon produced.

● ● ● ● ● ● ● ● ● ● ● ● ● ●

18.5   Summary

The layout of analog CMOS circuits is complicated by the fact that the operation can be very sensitive to small variations in the electrical parameters. This leads to the situation where it is not possible to provide a set of layout techniques that will work in every process.

Since analog circuits are a critical part in the evolution of CMOS VLSI, much effort has been put into the problem of physical design. Each group tends to have its own set of recommended geometries and techniques based on experience and feedback from the process line.

In this chapter, we have seen some of the main ideas in analog layout. The interested reader is directed to the excellent books listed in the References for a deeper treatment of analog CMOS ICs.

● ● ● ● ● ● ● ● ● ● ● ● ● ●

18.6   References

[18.1] Allen, P. E., and Holberg, D. R., *CMOS Analog Circuit Design, Second Edition*. New York: Oxford University Press, 2002.

[18.2] Clein, D., *CMOS IC Layout*. Woburn, MA: Newnes, 2000.

[18.3] Hastings, A., *The Art of Analog Layout*. Upper Saddle River, NJ: Prentice Hall, 2000.

[18.4] Razavi, B., *Design of Analog CMOS Integrated Circuits*. New York: McGraw-Hill, 2001.

● ● ● ● ● ● ● ● ● ● ● ● ● ●

18.7   Exercises

18.1   What is the gain of the circuit in the figure when the circuit is unloaded?

Problem 18.1

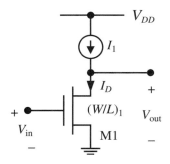

18.2   What is the gain of the circuit if loaded with a 100 Ω?

18.3   Layout the circuit in the figure. Determine the gain of the circuit conducting a simulation.

Problem 18.3

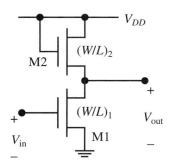

Microwind Command Summary

This is a summary of the Menu options available in Microwind. The latest available Microwind version has been included with this text. Upgrades are available on the Microwind website. Please see the *readme.doc* on the CD for the latest installation instructions, and Chapter 1, Section 1.5.2 Versions and Updates for URL.

● ● ● ● ● ● ● ● ● ● ● ● ● ●

A.1 File

This group of commands deals with all file operations.

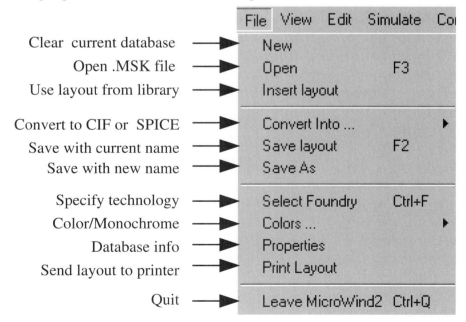

	File View Edit Simulate Co
Clear current database ➤	New
Open .MSK file ➤	Open F3
Use layout from library ➤	Insert layout
Convert to CIF or SPICE ➤	Convert Into ... ▶
Save with current name ➤	Save layout F2
Save with new name ➤	Save As
Specify technology ➤	Select Foundry Ctrl+F
Color/Monochrome ➤	Colors ... ▶
Database info ➤	Properties
Send layout to printer ➤	Print Layout
Quit ➤	Leave MicroWind2 Ctrl+Q

The sub-menus are shown below. These are self explanatory.

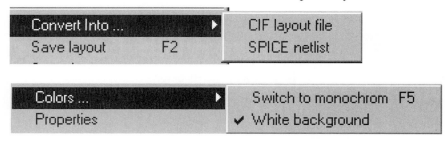

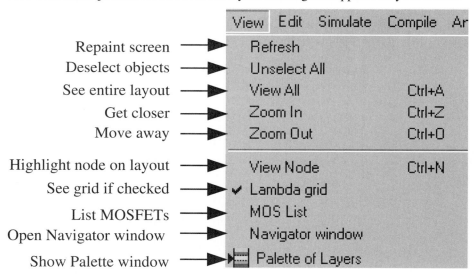

A.2 View

View commands provide control over the layout drawing that appears on your screen.

Repaint screen ➤ Refresh

Deselect objects ➤ Unselect All

See entire layout ➤ View All   Ctrl+A

Get closer ➤ Zoom In   Ctrl+Z

Move away ➤ Zoom Out   Ctrl+O

Highlight node on layout ➤ View Node   Ctrl+N

See grid if checked ➤ ✔ Lambda grid

List MOSFETs ➤ MOS List

Open Navigator window ➤ Navigator window

Show Palette window ➤ Palette of Layers

A.3   Edit

The Edit commands are the most important set for drawing. These allow you to do basic object manipulations and custom placement of all rectangles.

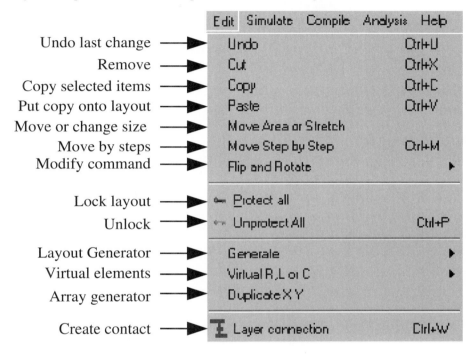

Undo last change → Undo — Ctrl+U
Remove → Cut — Ctrl+X
Copy selected items → Copy — Ctrl+C
Put copy onto layout → Paste — Ctrl+V
Move or change size → Move Area or Stretch
Move by steps → Move Step by Step — Ctrl+M
Modify command → Flip and Rotate ▶

Lock layout → Protect all
Unlock → Unprotect All — Ctrl+P

Layout Generator → Generate ▶
Virtual elements → Virtual R,L or C ▶
Array generator → Duplicate X Y

Create contact → Layer connection — Ctrl+W

Sub-menus of the Edit menu include:

- **Flip and Rotate**—Group objects, then perform an operation on the entire set

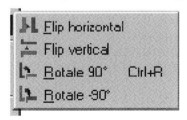

Flip horizontal
Flip vertical
Rotate 90°   Ctrl+R
Rotate -90°

- **Virtual R, L, or C**—Create electrical components for use in simulations. The elements are not part of the layout, but help you in the overall design

- **Generate**—This provides you with several pre-defined macros for transistors, contacts, wiring tools, and other electronic elements

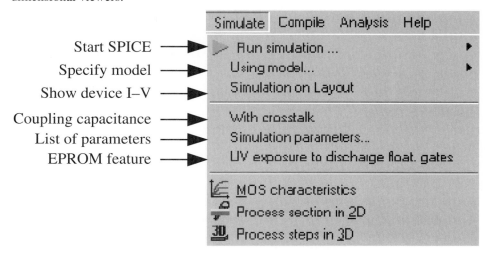

A.4   Simulate

This group of commands governs the SPICE simulations and the 2-dimensional and 3-dimensional viewers.

Start SPICE ⟶
Specify model ⟶
Show device I–V ⟶
Coupling capacitance ⟶
List of parameters ⟶
EPROM feature ⟶

The sub-menus are:

- **Run simulation**—Allows you to specify the analysis desired

Voltage vs. Time (Default)   Ctrl+S
Voltage, Current vs. Time
Static Voltage vs. Voltage
Frequency vs. Time

- **Using Model**—Select the SPICE MOSFET model to be used

Level 1
Level 3 (default)
BSim4 (advanced)

A.5 Compile

These two commands are for automated design.

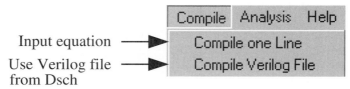

Input equation ⟶

Use Verilog file ⟶
from Dsch

A.6 Analysis

The Analysis menu invokes the Design Rule Checker (DRC) and the Measurement tool. In some versions, it also allows one to execute a parametric analysis.

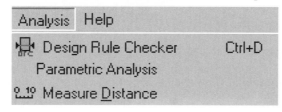

A.7 Help

This provides access to the online versions of the design rules and the reference manual. As mentioned earlier in Chapter 5, this book uses mostly generic SCMOS design rules. You must check current foundry design rules for updates.

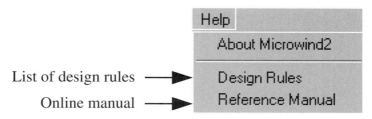

List of design rules ⟶

Online manual ⟶

A.8   Menu Bar

The Menu bar provides shortcut buttons for all major operations. Almost all also are accessible from sub-menus or keyboard commands.

The Menu bar buttons on the left side of the screen are summarized below.

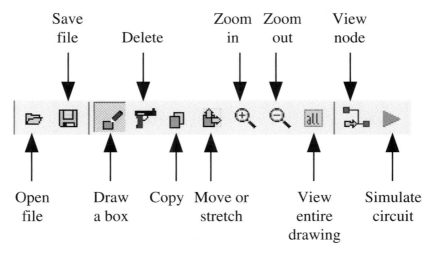

The buttons on the right side of the Menu bar are shown below.

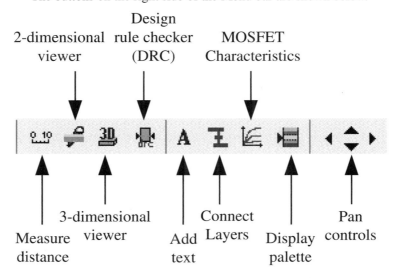

● ● ● ● ● ● ● ● ● ● ● ● ● ● ● ● ●

A.9   Other Screens

All major Microwind dialog and command windows are shown in the text under specific topical headings.

The online Help function and Microwind/Dsch User's Manual (on the CD) should provide sufficient information on these commands.

Microwind CMOS Technology Files

Microwind identifies technologies using the parameters that are defined in *.rul files*. These are text files and can be found within the Microwind folder.

The listing in this section is from the *cmos018.rul* file and is reproduced here to illustrate the details. The asterisk (*) is used for comment lines that make it easy to ready the data. This is the same design rule set that was discussed in Chapter 5. You can access other files directly from the CD. The first set of listings gives the geometrical design rules. Interconnect resistance and capacitance values are next, and SPICE parameters complete the listing.

Microwind obtains all of the values it needs from the *.rul file*. This includes layer information, design rules, device parameters, parasitics, CIF layer definitions, and data for 2-dimensional and 3-dimensional views. You can create a custom technology file by copying the format to a new text document, then editing the design rules, process specifics, electrical data, and SPICE parameter values. Saving the file with a .rul extension will make it accessible from the

File $\Rightarrow$
    Select Foundry

command sequence; alternatively, you may use the key strokes **[Ctrl][F]**. Be sure to give it a unique name that will not overwrite an existing file.

* *

MICROWIND 2.0
*
* Rule File for CMOS 0.18 μm
* Date : 18 May 98 by Etienne Sicard
* Date : 27 April 99 By Etienne/Fabrice
*   04 Jan 00 smaller dT

```
*       19 Fev 00 STI, Niso, low leakage,
*   high VT
*        6 May 00 Fit nMOS models
*       10 Mar 01 vddh 3.3V
*       20 Apr 01 r303 for high volt gates
*       12 Jun 01 poly2
*
NAME CMOS 0.18µm - 6 Metal
*
lambda = 0.1     (Lambda is set to half the
* gate size)
metalLayers = 6  (Number of metal layers :
*6)
lowK = 4.0       (inter-metal oxide)
salicide = 1     (Enable salicide 1=enable
* 0= disable)
*
* Design rules associated to each layer
*
* Well (Gds2 level 1)
r101 = 10    (well width)
r102 = 11    (well spacing)
*
* Diffusion (N+ 16, P+ 17, active 2)
*
r201 = 4     (diffusion width)
r202 = 4     (diffusion spacing)
r203 = 6     (border of nwell on diffp)
r204 = 6     (nwell to next diffn)
*
* Poly
*
r301 = 2     (poly width)
r302 = 2     (gate length)
r303 = 4     (high voltage gate length)
r304 = 3     (poly spacing)
r305 = 1 (spacing poly and unrelated diff)
r306 = 4     (width of drain and source diff)
r307 = 2     (extra gate poly)
```

```
*
* Poly 2
*
r311 = 2    (poly2 width)
r312 = 2    (poly2 spacing)
*
* Contact
r401 = 2    (contact width)
r402 = 3    (contact spacing)
r403 = 2    (metal border for contact)
r404 = 2    (poly border for contact)
r405 = 2    (diff border for contact)
* metal
r501 = 3    (metal width)
r502 = 4    (metal spacing)
* via
r601 = 3    (Via width)
r602 = 4    (Spacing)
r603 = 0    (via/contact)
r604 = 2    (border of metal&metal2)
* metal 2 (27)
r701 = 3    (Metal 2 width)
r702 = 4    (spacing)
* via 2 (32)
r801 = 3    (Via width)
r802 = 4    (Spacing)
r804 = 2    (border of metal2&metal3)
* metal 3 (34)
r901 = 3    (width)
r902 = 4    (spacing)
* via 3 (35)
ra01 = 3    (Via width)
ra02 = 4    (Spacing)
ra04 = 2    (border of metal3&metal4)
* metal 4 (36)
rb01 = 3    (width)
rb02 = 4    (spacing)
* via 4 (52)
rc01 = 3    (Via width)
```

rc02 = 4   **(Spacing)**
rc04 = 2   **(border of metal4&metal5)**
* metal 5 (53)
rd01 = 8   **(width)**
rd02 = 8   **(spacing)**
* via 5 (xx)
re01 = 5   **(Via width)**
re02 = 5   **(Spacing)**
re04 = 2   **(border of metal5&metal6)**
* metal 6 (xx)
rd01 = 8   **(width)**
rd02 = 15   **(spacing)**
*
* Passivation nitride (31) and pad rules
*
rp01 = 800   **(Pad width)**
rp02 = 800   **(Pad spacing)**
rp03 = 40   **(Border of Vias)**
rp04 = 40   **(Border of metals)**
rp05 = 200   **(to unrelated active areas)**
*
* Option layer around MOS
*
ropt = 5
*
* Thickness of conductors for process
*aspect
* All in µm
*
* P++ epitaxial
thepi = 1.0
heepi = -4.0
* niso description
thnburried = 1.0
henburried = -3.0
* Shallow tretch isolation
thsti = 1.0
hesti = -1.0
* Poly

thpoly = 0.20
hepoly = 0.01
* Poly2
thp2 = 0.2
hep2 = 0.22
* contact
thco = 0.7
heco = 0.5
thdn = 0.4
thdp = 0.4
thnw = 2.0
thme = 0.6
heme = 1.3
thm2 = 0.6
hem2 = 2.8
thm3 = 0.6
hem3 = 4.4
thm4 = 0.6
hem4 = 6.1
thm5 = 1.0
hem5 = 7.7
thm6 = 1.0
hem6 = 9.6
* Passivation
thpass = 0.5
hepass = 10.6
* Nitride
thnit = 0.6
henit = 11.2
*
* Resistances Copper
* Unit is ohm/square
*
repo = 4
reco = 2
reme = 0.15
revi = 1
rem2 = 0.06
rev2 = 2

```
rem3 = 0.06
rev3 = 23
rem4 = 0.06
rev4 = 1
rem5 = 0.03
rev5 = 1
rem6 = 0.03
*
*
* Parasitic capacitances
*
cpoOxyde= 4600 (Surface capacitance *Poly/Thin oxyde aF/µm2)
cpobody = 80   (Poly/Body)
cmebody = 28
cmelineic = 42
cmepoly = 60
cm2body = 20
cm2lineic = 30
cm2metal = 38
cm3body = 20
cm3lineic = 30
cm4body = 20
cm4lineic = 30
cm5body = 20
cm5lineic = 40
cm6body = 20
cm6lineic = 40
cgsn = 500 (Gate/source capa of nMOS)
cgsp = 500
*
* Vertical crosstalk
*
cm2me = 50
cm3m2 = 50
cm4m3 = 50
cm5m4 = 50
cm6m5 = 50
*
* Lateral Crosstalk
```

```
*
cmextk = 20     (Lineic capacitance for crosstalk coupling in aF/µm)
cm2xtk = 22     (C is computed using Cx=cmextk*l/spacing)
cm3xtk = 25
cm4xtk = 25
cm5xtk = 25
cm6xtk = 25
*
* Junction capacitances
*
cdnpwell = 350  (n+/psub)
cdpnwell = 300  (p+/nwell)
cnwell = 250    (nwell/psub)
cpwell = 100    (pwell/nsub)
cldn = 100      (Lineic capacitance N+/P-*aF/µm)
cldp = 100      (Idem for P+/N-)
*
*
* MOS definition
*
MOS1 low leakage
MOS2 high speed
MOS3 high voltage
*
* Nmos Model 3 parameters
*
NMOS
l3vto = 0.5
l3v2to = 0.34
l3v3to = 0.7
l3vmax = 100e3
l3gamma = 0.35
l3theta = 0.2
l3kappa = 0.08
l3phi = 0.5
l3ld = -0.02
l3u0 = 0.038
l3tox = 4e-9
l3nss = 0.04
```

```
*
* high speed
l3v2to = 0.35
l3u2 = 0.04
l3t2ox = 4e-9
*
* high voltage
l3v3to = 0.6
l3u3 = 0.06
l3t3ox = 8e-9
*
*
* Pmos Model 3
*
PMOS
l3vto = -0.6
l3v2to = -0.40
l3v3to = -0.70
l3vmax = 100e3
l3gamma = 0.4
l3theta = 0.3
l3kappa = 0.01
l3phi = 0.2
l3ld = 0.01
l3u2 = 0.02
l3t2ox = 4e-9
l3nss = 0.07
*
* high speed
l3v2to = -0.3
l3u2 = 0.02
l3t2ox = 4e-9
*
* high voltage
l3v3to = -0.7
l3u3 = 0.02
l3t3ox = 8e-9
*
* BSIM4 parameters
```

* Nmos
*
* Low leakage
NMOS
b4vtho = 0.55
b4k1 = 0.17
b4k2 = 0.1
b4xj = 1.7e-7
b4nfact = 1.69
b4toxe = 4e-9
b4ndep = 1.8e17
b4d0vt = 2.3
b4d1vt = 0.54
b4vfb = -0.9
b4u0 = 0.038
b4ua = 2.8e-15
b4uc = -0.047e-15
b4vsat = 80e3
b4pscbe1 =320e6
b4ute = -1.8
b4kt1 = -0.06
b4lint = -0.01e-6
b4wint = 0.02e-6
b4xj = 1.5e-7
b4ndep = 1.7e17
b4pclm = 0.29
*
* high speed
b4v2to = 0.4
b4t2ox = 4e-9
*
* high voltage
b4v3to = 0.7
b4t3ox = 8e-9

*
* Pmos BSIM4
*
PMOS

```
b4vtho = 0.5
b4k1 = 0.29
b4k2 = 0.1
b4xj = 1.7e-7
b4nfact = 2.0
b4toxe = 4e-9
b4ndep = 1.8e17
b4d0vt = 2.3
b4d1vt = 0.54
b4vfb = -0.9
b4nfact = 2.2
b4u0 = 0.01
b4ua = 1e-15
b4uc = -0.047e-15
b4vsat = 60e3
b4pscbe1 =320e6
b4ute = -1.8
b4kt1 = -0.06
b4lint = -0.04e-6
b4wint = 0.02e-6
b4xj = 1.5e-7
b4ndep = 1.7e17
b4pclm = 0.3
*
* high speed
b4v2to = 0.4
b4t2ox = 4e-9
*
* high voltage
b4v3to = 0.7
b4t3ox = 8e-9
*
* CIF Layers
* MicroWind layer, CIF layer, overetch
*
cif nwell 1 0.0
cif diffp 17  0.1
cif diffn 16  0.1
cif aarea 2  0.0
```

```
cif poly 13 0.0
cif poly2 14  0.0
cif contact 19  0.025
cif metal 23 0.0125
cif via 25  0.0125
cif metal2  27 0.0125
cif via2 32 0.0125
cif metal3 34 0.0125
cif via3  35 0.0125
cif metal4 36 0.0125
cif via4  52 0.0125
cif metal5 53 0.0
cif via5  54 0.0
cif metal6  55 0.0
cif passiv 31 0.0
cif option  40 0.0
cif text 94 0.0
*
*
* MicroWind simulation parameters
*
deltaT = 0.5e-12   (Minimum simulation * interval dT)
vdd = 2.0
hvdd = 3.3
temperature = 27
riseTime = 0.05
*
* End CMOS 0.18 µm
*
*
```

Index